Traveling Weatherwise in the U. S. A.

Edward
Powers
and
James
Witt

TRAVELING WEATHERWISE IN THE U.S.A.

With 150 weather maps, charts, and tables

DODD, MEAD & COMPANY   *New York*

ISBN: 0-396-06360-8

Library of Congress Catalog Card Number: 73-153892

Printed in the United States of America

To Bert, who left her footprints on almost every spot mentioned,
and Nancy, who enjoyed vicarious journeys while editing this volume

Foreword

After many years of traveling throughout the world and a careful consultation with whatever guide books we could find, there was one recurring factor which continually seemed to interfere with our plans. We could locate the most desirable hotels and restaurants and take advantage of off-season rates and charter travel, but outside of "average" temperature forecasts, there was no available guide to the local weather conditions.

What we have tried to do in *Traveling Weatherwise in the U.S.A.* is to assemble just that guide in a ready reference form. It is intended neither as a travel guide nor as a meteorological study. There are no isobars or pressure systems, nor will you find a rating of hotels. It is simply intended to provide a layman's eye view of the seasonal variations within the twelve national weather areas in the United States, and specifically within over 100 American cities on a month-to-month basis.

Traveling Weatherwise in the U.S.A. is intended for the traveler on business or pleasure, and for the retiree looking for the best potential retirement residence. You will find as we did that there are many opportunities for combining the lowest off-season rates with some of the best weather a given location has to offer. We hope we have been able to provide, at least to some extent, a useful and comprehensive guide to the weather of the United States.

EDWARD POWERS
JAMES WITT

Contents

CONTENTS

Charts and Tables

xi

CHARTS AND TABLES

7 WHAT IS WEATHER?

Introduction

"Have a good time? How was the weather?" These are the usual first greetings—unless of course, the returning vacationist or traveler is well tanned and then it will probably be, "Well, I see you had lots of sunshine."

Weather may not be the most important subject in the world, but it is said to be the chief reason for the second "hot line" between Washington and Moscow and remember Gemini V had to return earlier than planned—because of an oncoming storm. Mark Twain wasn't the first to complain that no one did anything about it.

To be sure, we have tried and can now duplicate Mother Nature and produce a perfectly controlled climate—but, alas, only indoors! No one can yet guarantee even one sunny day. Despite the odds, however, there is an answer. A little previous homework can greatly improve the chances of enjoying pleasant conditions by selecting the most favorable times and locations.

But where can one find this information? Well—there are many scholarly tomes devoted to climate and weather but, unfortunately, almost all are written from the technical viewpoint of the meteorologist and are of little practical use to the tourist or retiree seeking a comfortable spot to cast anchor.

And that, dear reader, is the primary excuse for this book.

No, it is not a new and better travel guide. There are already plenty and many of excellent quality. All are filled with useful information—places, sights and events—how to get there and back, where to stay, what to eat and, of course, drink. But with so much to cover, it is understandable that this important aspect of physical comfort must get rather brief coverage. Some settle for a paragraph on "what to wear," while others include abbreviated tabulations of temperature and precipitation for a single city in each country or state. Almost always, information is in general terms and average figures, although there are a few outstanding exceptions.

Professors delight in explaining that a person can drown in water averaging three feet in depth. But so, too, can he either shiver or broil in an average temperature of 60 degrees. Statistics are a necessary evil, but if incomplete, they can also be very misleading. As for example: an average daily temperature of say 70 degrees may mean either uniform comfort throughout the twenty-four hours, or a sizzling 95 degrees midday, with a chilling 40 degree drop to evening goose pimples. Either climate may be the one you prefer—but it's nice to have the choice.

Now, if all of this is so important, why has it been so neglected when there seem to be mountains of books on every conceivable subject from aardvarks to zymosis? As usual, there are sound reasons.

Until quite recently, travel habits had generally followed rather long-established patterns of both time and place. Each vacation spot had its recognized season, during which time the weather was usually satisfactory—or at least better than at home. Consequently, there was little need for a book of this type.

But times have changed. The tremendously increased demand in recent years has packed available vacation space to bursting during the top season. Greatly liberalized vacation schedules, ever-increasing business travel and conventions, and great armies of more affluent retirees, often combine to make more people than space available. Resorts and transportation tend to oversell and reservations can often be booked only far, far in advance. That may not always be convenient or even possible and, in any case, you will almost surely end up paying the highest rates for the privilege of joining hectic crowds, ofttimes with hurried or sloppy service.

As a result, many are beginning to look hopefully further afield, but even with jet transportation and special rates, time and cost can limit distance. One good alternative is to investigate off-peak periods and new travel patterns. In many cases (but certainly not all), weather during the "in" season is neither the most pleasant nor desirable. To be sure, other periods do involve risk but with careful planning, the results can be more comfortable leisurely service, greater flexibility of itinerary and, most often, lower rates.

But all of these, of course, are not enough, if the weather is not good. To most of us each vacation or trip is a bit of an adventure and improving the chances of it's being an enjoyable experience with pleasant memories should be worth a little effort.

All travelers and vacationists, whether embarking on a "five-dollars-a-day" jaunt or booking a "presidential suite" tour, have one interest in common—the hope for pleasant weather.

It has been our aim to compile rather complete "travel-weather" information in a form which we do not believe has ever been tried before. The simple untechnical language may be more useful for the traveler, vacationist, and retiree who couldn't care less about dew point, relative humidity, etc.—just as long as it doesn't rain.

Taking a wild guess at the weather is perhaps a bit like putting a stack of chips on that "hunch" roulette number in Las Vegas. You may win but—the odds just ain't your way.

If this book is used in conjunction with a good travel guide and the experienced advice of your travel agent, it is our hope that your next travel adventure may be a little less of a gamble.

1

An Overall Look at the Country

First let's take an overall look at the climatic happenings throughout mainland United States (except Alaska) which may affect your comfort, health and pleasure.

We have generally tried to present much information in graphic form. However, because the conventional weather map has almost zero value to the average reader, we have used less technical, picture-type charts prepared with primarily the vacationist, traveler and retiree in mind.

People differ greatly in their reaction to weather. What may be of almost casual moment to one, can be of serious concern to another—even causing someone to completely avoid a particular area. It may be anything from high pollen count to excessive humidity, thunderstorms or great earthquake potential.

By referring to this series of illustrations the reader can quickly form a composite picture of overall conditions throughout the country and pinpoint regions which may be of most interest. The retiree might be most interested in the number of clear days or hours of sunshine in the year—whereas the vacationist or traveler will want more precise information covering a more limited period of time.

After developing some general ideas about the climate of one or several areas of the country which may look attractive, the reader can then come to a definite conclusion by studying the more complete descriptions of each of those places in chapter 2. You will note from chart no. 16 that the country has been divided into twelve areas somewhat along climatic lines. It is easy to proceed from one to the next, examining each in complete detail on the way.

Chapter 2 is set up as a convenient ready reference handbook, so that a given city or area can be found without plowing through a great mass of reading matter. The climate of each of the twelve climate areas of the United States treated in this book is covered separately by seasons. There is also a larger scale map, a complete descriptive text, and tables showing the weather conditions in a scattering of cities across each of the areas.

In spite of the effort we have made towards simplicity and clarity, it is not easy for most of us to actually visualize such things as 42 inches per year of rainfall and even humidity figures and such may have little real meaning.

We will suggest here, and probably many times in these pages, that you develop the habit of constantly comparing the data given for any particular geographical section with that shown for your home area or a place with which you are familiar. This method will enable you to better visualize the weather conditions which you may reasonably expect to find.

The brief descriptions of the various types of storms will never make you an expert but they may help correct some erroneous or exaggerated impressions concerning them. As an example, there are many who would hesitate about living in Florida only because of hurricanes. A "Donna" or "Carol" may get such wide headline publicity, that great numbers of people avoid even visiting the state.

Others have a natural distaste or even dread of a particular type of weather disturbance whether it be thunderstorms, tornadoes or whatever. These little stories and maps will indicate the areas of greatest frequency and severity. It is surprising how many find foggy, windy or other weather conditions so very objectionable that they will by-pass places where they prevail.

While there is no "ideal" spot, a little effort and study here may help you determine several places offering the most plus values for your own taste. At least, for the moment, there seems to be no easier method of guessing ahead what the weather may be.

At a recent seminar, a scientist suggested that full year weather predictions may be just beyond the horizon—perhaps even by the end of this century. It was indicated that if several readings per minute from the six hundred stations around the world, plus some in China, could be fed simultaneously into a computer 150 times more powerful than any now available, it would require only a few billion calculations to turn out a forty-eight-hour prediction.

But perhaps the human element may be required for a while longer and unless you have the patience to wait, you may have to get along with this book for the next thirty years or so.

Climatologists divide the United States into areas based upon various considerations, including temperature, precipitation, and almost every other element relating to weather. Individual sections of the country are often designated as being generally similar to other parts of the world. An example might be the California coast, which is said to have a "mediterranean climate." Others use such descriptive terms as "humid sub-tropical." The United States Army has developed its own very complete set of charts showing climatic zones.

Such charts are usually very well prepared and each can contribute to the building up of a good overall picture. Simplified maps which are limited each to a single item such as thunderstorms, precipitation, etc., can also be very informative. There are a number of that type included in this chapter.

People often incorrectly use the words *weather* and *climate* interchangeably. Weather is, of course, the state of affairs at the moment. Climate is the aggregate of all weather; it is the composite or generally prevailing conditions of a region averaged over a series of years.

The United States includes so great a variety of weather conditions that there is almost no known climate which isn't closely duplicated somewhere within its boundaries.

There are large regional areas of weather which are quite easy to identify and remember. Within these vast climatic regions however there can be, and most often are, many small enclaves which experience weather conditions differing very greatly from the surrounding areas. To accept overall descriptions without further detailed investigation can result in great disappointment. For instance, there are portions of the California coast which experience frequent dense fogs. In some places the fog patterns are very distinct and predictable, which can be the case where the fog bank follows a valley or is affected by air currents. Spots only a short distance apart can vary considerably in frequency of fogs. One side or the other of a small mountain range may experience great differences of precipitation, particularly if one slope is exposed to moist air currents from the sea. In some high country sections, which receive a fair amount of winter sunshine but are subjected to high winds and blustery periods, there are small valleys which enjoy exceptionally mild and pleasant weather while the surrounding country shivers.

We mention this only to alert the reader to always be on the watch for conditions which could produce a small weather enclave or mini-climate. They are not easy to find. Unfortunately there is no tabulation nor single source from which even reasonably complete information can be obtained. By and large you will most often have to depend upon personal scouting fortified with some weather deductions. We have tried in this book to give information and suggestions which include hints on where and how to at least begin the search. Actually it need not be a haphazard undertaking, for there is a logical reason and explanation for every climatic condition.

Two important basic factors, of course, are distance from the equator and altitude. But there are many others that may have as much or even more influence on predominating characteristics. This is particularly the case in local areas. They include ocean and air currents, distance from large bodies of water, prevailing winds and, of course, the surrounding topographical features. It might be a valley, a hill, the shape of the coast line or even the side of the mountain you choose. Similar considerations govern weather conditions elsewhere in the world. What makes one hectare of French vineyard worth many times as much as others only a short distance away? It certainly isn't the soil alone. The key may include getting maximum or morning sunshine, protection from high winds and storms, or that bête noir of the French viniculturist,—hail. There might be any of a dozen more reasons, but most are related to weather.

So whether you are at the initial stage of trying to choose a general area or have reached the point of selecting a definite spot within it, the factors govering the decision are much the same. The big and very important difference is that the large climatic areas are well documented with complete and accurate information accumulated over a period of many years. The small ones are mostly do-it-yourself projects.

Because of the latitude, most of this country, has a humid continental climate. This indicates distinct seasonal changes with warm to hot humid summers and cool to cold winters, graduated from north to south, coupled with frequent and often rapid changes of weather. This generally describes the eastern five-eighths of the nation, except that the southern portion would better be classified as humid sub-tropical. This is the vast region which because of the high barricading mountain ranges along the east and west coasts is relatively unaffected by the oceans and the moisture laden westerly winds. It is, however, exposed to both the wintery blasts from the north and the warm humid breezes from the Gulf.

The extreme eastern coastal strip is somewhat sheltered from the most severe western storms by the lofty Appalachian ranges. When the wind shifts, however, there can be blustery or even violent northeast storms from the sea. As the prevailing winds are westerly, the Atlantic ocean has little influence on the land, although the winter temperature along the immediate coast may be 5° to 10° higher than even a few miles inland. This relatively small difference can have quite a bearing on the length of the growing season, and in extreme cases it may increase it as much as 40 days over the nearby inland countryside. Snowfall is also much lower along the coast than inland and at the higher elevations. Snow cover close to the sea is generally less; and south of Boston, the major portion of precipitation is in the form of rain.

Thunderstorms are quite common in the East, particularly in Florida and along the Gulf coast. Perhaps to compensate, coastal parts get somewhat more winter sunshine than the interior, but northern winters by and large are associated with overcast skies. Autumns are almost always glorious and spring, when there is one, can be a delight. They are by far the most pleasant seasons in these parts.

The Gulf Stream influences the weather along the extreme southern coasts but turns eastward before it has any pronounced effect on most of our other shores. Labeled by many appreciative Europeans as "America's finest export," it causes cactus and palms to flourish on the southwest corner of England.

The Gulf country is one of the wet belts, averaging about 60 inches of precipitation a year, with no really dry periods. Because there is no protecting chain of mountains extending east and west, the cold blasts from the north, and even Canada, do on occasion surge down to mid-Florida, bringing along freezing wintery weather.

Humidity is high and can be intolerable to many during the hot Gulf summers. Florida, which has the nation's highest thunderstorm frequency (70 to 90 per year), also gets heavy rainfall—but little of it during the sunny winter season. The Gulf and Atlantic coasts are also the targets of the occasional hurricane which strays inland.

All the ranges of the long Appalachian chain are to a degree wet, green mountains. The northern portions provide some of the finest skiing in the nation and this entire high country is a very popular summer playground. The Blue Ridge section in particular has frequent heavy fogs which range from 30 to 50 days a year.

The Great Lakes region is one of winter turbulence with the wind, and sometimes the storms, blowing in from almost any direction. Such large water bodies affect surrounding lands. In this case they lengthen the growing season appreciably, which accounts for the great abundance of fruit, grapes and berries grown along their shores, particularly on the southern and eastern sides. The average lakeside winter temperature is perhaps 5° higher than the country a few miles inland. There is also a slight lowering of the summer temperatures as compared to the surrounding areas. While there is not too much effect on the total precipitation, snowfall is increased substantially especially in the famous "snow belt" sections. Thunderstorms are fewer close to the lakes but fog is heavier—particularly south of Lake Superior, which may experience 25–30 days of dense fog a year. Most of this region enjoys a good sunshine average, but overcast skies are most common in winter with less sunshine than might be expected for this latitude.

The western third of the country, as the topography would suggest, is rather dry. The moisture-laden westerly sea breezes are effectively barricaded by the many ranges of lofty mountains extending from the Canadian to the Mexican border. The northern half of this area can experience exceptionally rugged winters as this is real blizzard country. Summers are hot and sunny. The whole north and south central portion of the United States is the birth place of most tornadoes. Further to the south, centering in Arizona and New Mexico, is classified as tropical desert. It is a land of golden sunshine (up to 90% of the maximum possible in the Yuma area), with little time for rain. Nevada is the driest state, averaging less than 9 inches total precipitation a year. Summer temperatures can top 100° with great regularity. Refugees from these parts escape, when possible, to the comforting breezes of the Pacific. Winter and particularly spring are magnificent.

The strip between the Pacific Ocean and the mountains is unique and differs markedly from the large inland valleys and lands east of the mountain ranges. The California Current and the prevailing westerly winds temper the climate along the immediate coast, greatly reducing the temperature extremes. There is plenty of fog, which man all too often transforms into the dreaded smog. Point Reyes, just north of San Francisco, is about the foggiest spot in the whole country. Most of the winter precipitation is in the form of rain or, at most, very wet snow which doesn't remain long on the ground. The Oregon coastal area is usually said to have a marine climate, while that to the south is designated as mediterranean. The southern end enjoys considerably more sunshine, particularly in winter when the northern coast is almost continuously overcast, damp, and dreary. Summers along the Washington and Oregon coasts, however, are almost perfect. Many claim that the San Diego area of southern California has the finest year-round weather of any place in the U.S.A.

The relative uniformity of the temperature along the Pacific coast is due in no small part to the California Current, which is the only cool summer water in this vast ocean of warm tropical streams. It flows south along the coast from about March through July, cooling the air which moderates the summer extremes. This unusual natural phenomenon is caused by a combination of the earth's rotation and action of the wind which forces the surface water near the shore out into the ocean. The cold water from below is sucked up to fill the void; this is called "upwelling." The sequence is reversed in the fall.

This is a very simplified description of the overall weather patterns of the forty-eight contiguous states. Used in conjunction with the thumbnail maps, however, the total information should help guide the reader to decide in which section he may have the greatest interest. From that point, he can follow up specific areas in detail in chapter 2.

The mini or sub-climates are the result of much the same conditions that influence large areas. As an illustration, one can see how effectively the high, western mountain ranges barricade the moisture-laden air from the Pacific and wring it almost dry before it passes into the inland valleys and the country beyond. The results can be seen quite dramatically from a plane window. The west side of the mountains may be a lush green while the eastern slopes appear dry and brownish.

Many people are attracted to the northwest country by the magnificent summers and general atmosphere; the dreary winter days, however, can change their minds. While the weather is not at all severe during that season, the coastal lands north from about San Francisco can be cool, damp, foggy, and in some stretches quite windy. Actually it is very similiar to the winter climate of southwestern England.

At first glance, the only alternative for remaining in the general area would seem to be the wide inland valleys just on the eastern lee of the high mountains. While these lands escape much of the fog, wind and dampness, and even have a bit more sunshine, the winters can be extremely cold and the summers real scorchers.

An attractive compromise might be a spot like Stockton, which is east of San Francisco. It's located right in the middle of the Great Valley about at the junction of the Sacramento and San Joaquin rivers. The redeeming feature is that it is almost opposite a wide opening in the mountains through which the river flows toward the Pacific. The prevailing ocean breezes drift east through the gap, moderating both the hot summer and low winter temperatures. Yet it is still close enough to enjoy all that the coast has to offer.

After reading about the overall weather characteristics of an area, it is recommended that you take a look at the local topography, air and sea currents, and any other factor that may affect the particular spot that you have in mind. This is especially important if you contemplate living there or plan a long stay. It would surprise many to hear that Miami, on the east coast of Florida, averages about 18 days of 90° temperature per summer but that Fort Meyers, on Florida's west coast, may have 100 or more such days. Or that one place about 50 miles from another may have almost 10 times as much precipitation a year.

In many cases, you may not have sufficient information to arrive at a very accurate conclusion. By and large, because of space requirements, the weather information in guide books and travel literature includes only temperature and precipitation averages. Such limited data can be misleading. As an example, Seattle—known as being in "the land of the web foot"—gets a total annual precipitation of 33.4 inches while sunny Miami averages a fat 59.7 inches per year. This is confusing only because it is not the whole story.

Among the charts, tabulations and text we have included the following information regarding the weather in these two cities:

	Seattle	Miami
Total annual precipitation in inches	33.4	59.7
Average annual temperature range	59°-45°	81°-69°
Total clear days per year	80	100
Total cloudy days per year	160	120
Total foggy days per year	53	6
Total number of days with 0.01" or more of precipitation per year	163	128
Total thundershowers per year	8	80
Per cent of maximum possible sunshine	45	67
Total hours of sunshine per year	2019	2903

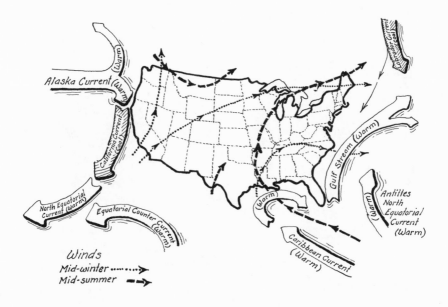

Winds
Mid-winter ....➤
Mid-summer ----➤

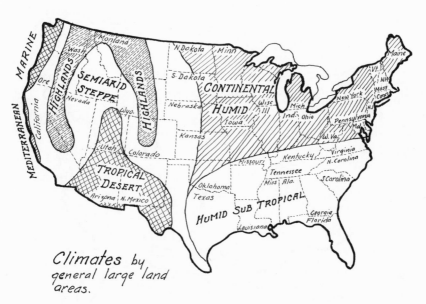

Climates by
general large land
areas.

CLIMATES & CURRENTS

Chart No.1

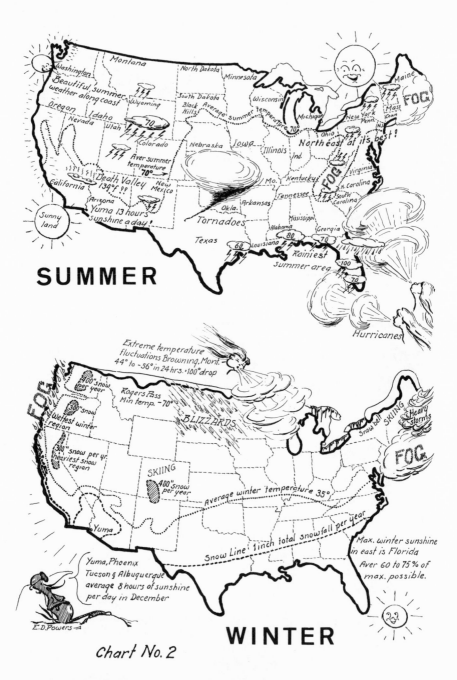

SUMMER

Montana · North Dakota · Minnesota

Washington
Beautiful summer weather along coast

Wyoming · South Dakota · Black Hills · Wisconsin · Michigan · Maine · FOG · Mass · Conn · New York · Penn · N.J.

Oregon · Idaho · Utah · 70 · Average summer temperature 70° · Ohio · Northeast at its best!

Nevada · Colorado · Nebraska · Iowa · Illinois · Ind. · Virginia

Aver summer temperature 70°

California · Death Valley 134°f ?? · New Mexico · Mo. · Kentucky · N. Carolina · FOG · South Carolina · Tennessee

Arizona · Yuma 13 hours sunshine a day! · Okla. · Arkansas · Mississippi · Alabama · 80 · Georgia · 70

Sunny land

Tornadoes · Texas · 60 · Louisiana · 70 · Rainiest Summer area · 100 · 70

Hurricanes

Extreme temperature fluctuations Browning, Mont. 44° to -56° in 24 hrs. = 100° drop

FOG

400" snow per year · Rogers Pass Min temp. -70° · BLIZZARDS · Snow belt · SKIING · Heavy storms

300" snow · Wettest winter region

300" snow per yr. heaviest snow region · SKIING · 400" snow per year · FOG

Average winter temperature 35°

Yuma

Snow Line 1 inch total snowfall per year

Yuma, Phoenix Tucson & Albuquerque average 8 hours of sunshine per day in December

Max. winter sunshine in east is Florida Aver 60 to 75% of max. possible.

E.D.Powers

WINTER

Chart No. 2

STORMS & SUNSHINE

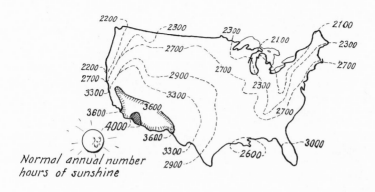

Normal annual number
hours of sunshine

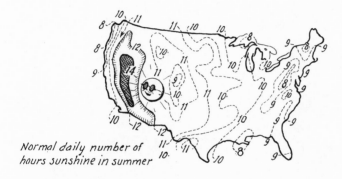

Normal daily number of
hours sunshine in summer

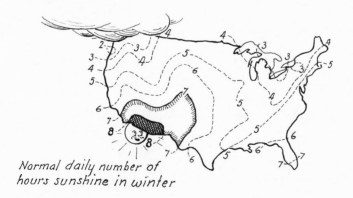

Normal daily number of
hours sunshine in winter

HOURS OF SUNSHINE

Chart No.3

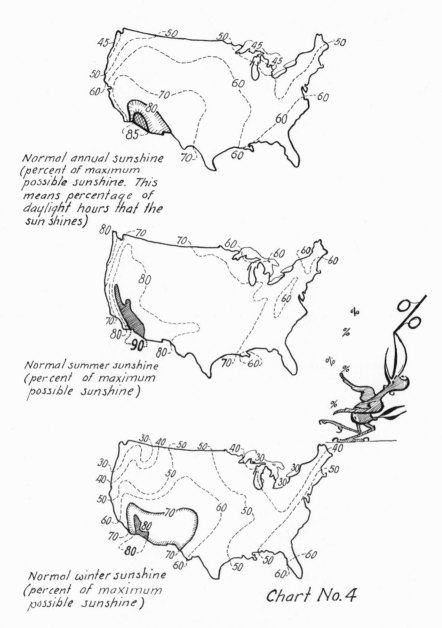

Normal annual sunshine
(percent of maximum
possible sunshine. This
means percentage of
daylight hours that the
sun shines)

Normal summer sunshine
(percent of maximum
possible sunshine)

Normal winter sunshine
(percent of maximum
possible sunshine)

Chart No. 4

SUNSHINE
PERCENT OF POSSIBLE

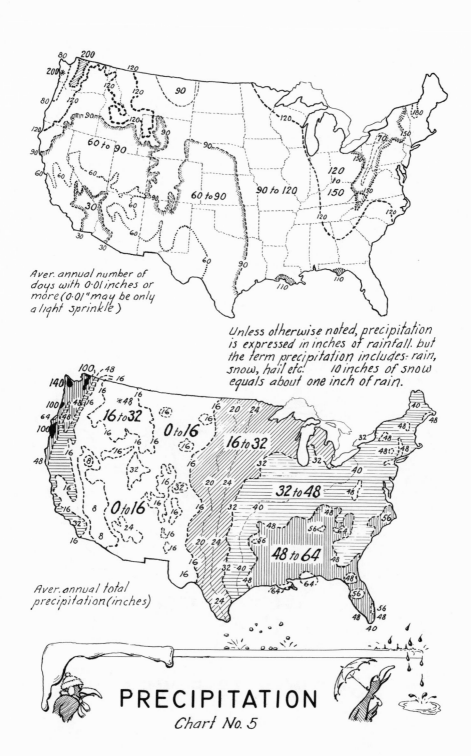

Aver. annual number of
days with 0·01 inches or
more (0·01" may be only
a light sprinkle)

Unless otherwise noted, precipitation
is expressed in inches of rainfall. but
the term precipitation includes: rain,
snow, hail etc.   10 inches of snow
equals about one inch of rain.

Aver. annual total
precipitation (inches)

PRECIPITATION
Chart No. 5

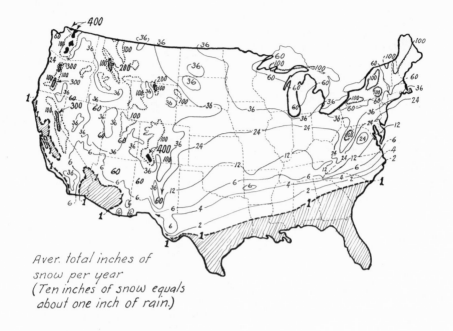

Aver. total inches of
snow per year
(Ten inches of snow equals
about one inch of rain.)

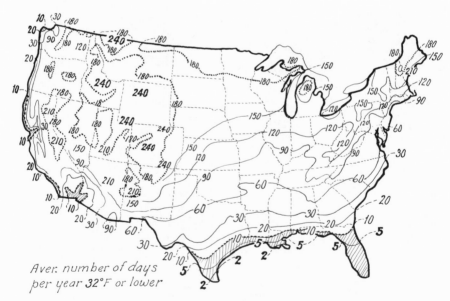

Aver. number of days
per year 32°F or lower

SNOW & 32°F WEATHER

Chart No. 6

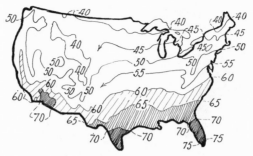

Average annual temperature

Average summer temperature

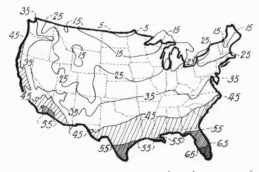

Average winter temperature

AVERAGE TEMPERATURES

Chart No. 7

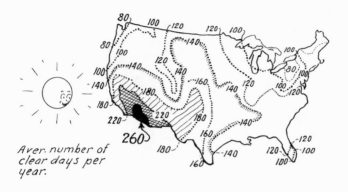

Aver. number of
clear days per
year.

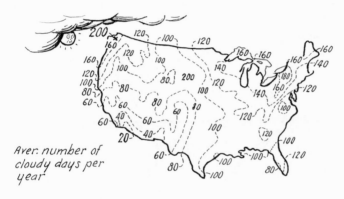

Aver. number of
cloudy days per
year

Aver. number of
dense fog days
per year

SUN OR CLOUD CONDITIONS

Chart No. 8

Highest temperatures (°F)
experienced in normal year
▨ = 100°F areas
Hottest spot in U·S·A·
134° Death Valley, Calf., July 10, 1914.

Lowest temperatures (°F)
experienced in normal year
▨ = 0°F areas
Coldest spot in these 48 States
-69·7°F, Rodgers Pass, Mont.

Average surface wind
velocity (miles per hour)
▨ = 8 M·P·H· areas
World's highest surface velocity
225 M·P·H· Mt Washington, N·H·

Chart No. 9

ZERO, 100°F
& WIND

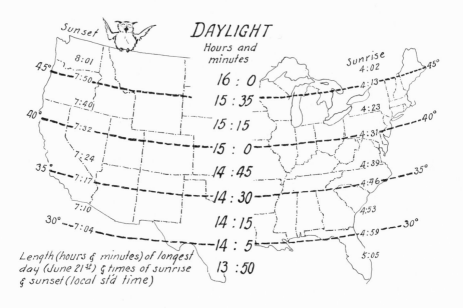

Sunset

DAYLIGHT
Hours and minutes

Sunrise
4:02

45° 8:01 .... 4:13 45°
7:50 16 : 0 4:13

7:40 15 : 35 4:23
40° 15 : 15 40°
7:32 15 : 0 4:31

7:24 14 : 45 4:39
35° 14 : 30 35°
7:17 4:46

7:10 14 : 15 4:53
30° 7:04 14 : 5 4:59 30°

5:05

Length (hours & minutes) of longest day (June 21st) & times of sunrise & sunset (local std time)

13 : 50

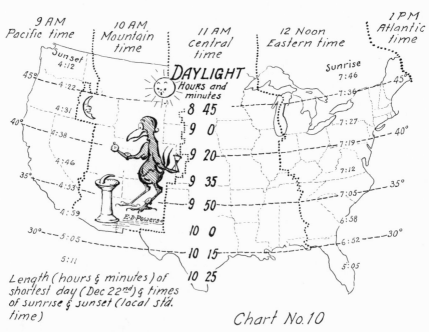

9 AM
Pacific time   10 AM
Mountain
time   11 AM
Central
time   12 Noon
Eastern time   1 PM
Atlantic
time

Sunset
4:12   Sunrise
7:46

45° 4:22 ## DAYLIGHT 7:36 45°
Hours and
minutes

40° 4:31 8 45 7:27 40°
4:38 9 0 7:19

4:46 9 20 7:12
35° 4:53 9 35 7:05 35°

E.D.Powers 9 50 6:58
4:59

30° 5:05 10 0 6:52 30°

5:11 10 15 5:05

Length (hours & minutes) of shortest day (Dec 22nd) & times of sunrise & sunset (local std. time)

10 25

Chart No.10

DAYLIGHT HOURS

This may be going a bit overboard in giving information, but a quick comparison of the figures should give a fairly good picture of the weather conditions that you might reasonably expect to find in the two cities. The above, of course, are only the annual averages; there are also monthly figures and descriptive texts of each area.

This has been a quick skirmish through the large climatic divisions of the forty-eight states, with an overall look at the particular conditions which prevail. The next step will carry us on to the much more detailed information in chapter 2.

THUNDERSTORMS

See Chart No. 11

An oversimplification of why lightning and thunder occur might be:

Lightning is a gigantic electric spark between areas where positive and negative electrical charges become concentrated. The phenomenon is similar to what we call *static electricity,* which causes the spark that may occur as a person touches metal after scuffing his feet on a carpet.

The thunder, which accompanies lightning, is generated by the very rapid expansion of the air in the path of the lightning stroke. This expansion results from the sudden heating of the air by the lightning.

It is variously estimated that 300 to 500 people are killed each year in the United States by lightning. Although thunderstorms may occur at any hour of the day, the greatest frequency is during late afternoon and evening. The seasonal activity generally peaks in July or August; but as indicated on chart no. 11, it may take place in May or June.

Because the percentage of thunderstorms is so high during warm weather, the summer frequency map greatly resembles that for the full year. The winter frequency map shows quite a different pattern, with little or no storm activity in the western half of the country. There are the usual transition periods between. A narrow strip extends along the Pacific Coast where the thunderstorm score is almost zero.

Many are curious about estimating the distance away from lightning. For rough approximation, divide the number of seconds between the sight of the lightning and the sound of thunder by five. The resulting

19

Month normally having the most
thunderstorms

Average number of thunderstorms per
year. (summer pattern is quite
similar)

Normal number of
thunderstorms in
winter.

E.D.Powers

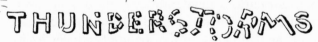

THUNDERSTORMS

Chart No.11

figure will be the approximate distance in miles between you and the lightning.

The ordinary precautions that should be heeded are issued by safety councils and posted in many recreational areas. It is probable that many of the lightning deaths might be avoided if reasonable observation of these suggestions were observed.

HURRICANES

See Chart No.12

In Japan and the Pacific generally, these storms are called *typhoons;* in the Bay of Bengal and Arabian Sea, *cyclones;* while in Australia they are known as *willy-willies.* That kind of destructive storm by any other name, however, is still a *tropical cyclone,* just as are our hurricanes.

But there are other types of violent disturbances. The tornado blows with more concentrated fury but is quite local in character and its path rarely exceeds 1000 feet in width. Temperate zone storms from off the ocean are usually larger, but the hurricane is by far the most dangerous and destructive.

Starting as tropical storms, they do not become hurricanes unless the whirling air reaches a velocity of 74 mph, which is a reading of 12 on the Beaufort scale. These storms originate in the tropical oceans, never on land. Whirling counter-clockwise (they revolve clock-wise in the Southern hemisphere), they churn up mountainous waves, sometimes causing flood tides up to twenty-five feet. Long coastal strips have been inundated and whole towns wiped out—indeed, the major damage is caused by water rather than wind.

The annual total of hurricanes striking land along the Atlantic and Gulf coasts has varied from 1 or 2 to the record 21 in 1933, but all of these did not cause heavy damage. The 3700 miles of coastline from Eastport, Maine to Brownsville, Texas is a wide target and the chances of a hit at any one point are very small. Summer is the season of greatest activity, and usually August, September and October account for over 80% of the annual total.

In spite of the great publicity, it can be seen from chart no. 12 that the frequency is extremely low and the chances of an individual actually meeting up with one of these storms is rather remote and each year the

21

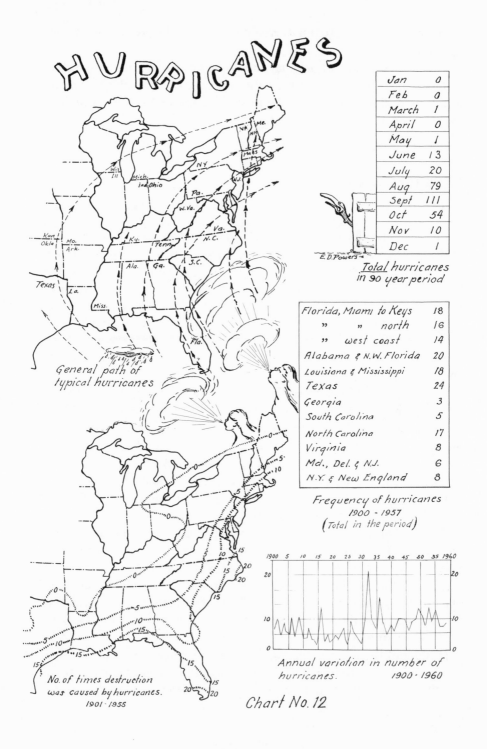

HURRICANES

Jan	0
Feb	0
March	1
April	0
May	1
June	13
July	20
Aug	79
Sept	111
Oct	54
Nov	10
Dec	1

E.D.Powers

Total hurricanes in 90 year period

Florida, Miami to Keys	18
" " north	16
" " west coast	14
Alabama & N.W. Florida	20
Louisiana & Mississippi	18
Texas	24
Georgia	3
South Carolina	5
North Carolina	17
Virginia	8
Md., Del. & N.J.	6
N.Y. & New England	8

Frequency of hurricanes
1900 - 1957
(Total in the period)

General path of typical hurricanes

No. of times destruction was caused by hurricanes.
1901 - 1955

Annual variation in number of hurricanes.   1900 - 1960

Chart No. 12

danger of personal injury decreases. The National Hurricane Center in Miami and other government agencies do a magnificent job in tracking hurricanes and broadcasting warnings. If the current program is successful, there will soon be a 15 to 18-hour lead time alert, including 12 hours of daylight. While property damage may continue, it is reassuring to know that the personal hazard has been greatly removed from hurricanes.

Some observers believe that the overall pattern of hurricane paths changes over long periods of time, perhaps in cycles. The Environmental Science Services Administration has plotted the courses of devastating (those causing damage) hurricanes since the beginning of this century. The results suggest that such changes do take place. There was a total of eight such hurricanes from 1900 to 1928. Seven centered in the western Gulf coast and only one traveled up the Atlantic side. Of the eight from 1935 thru 1945, all but one went up the Atlantic coast. The eleven in the 1947–54 period were scattered but centered on Florida. The thirteen from 1955 to 1965 were also widespread, with five in the Gulf and eight on the Atlantic side. Some think that the trend is again in the Gulf direction, but the experts seem to be withholding judgment. It is good to note that while property damage remains high, the loss of life has steadily decreased as a result of the ever-improving warning systems.

TORNADOES

See Chart No. 13

Tornadoes are certainly the wildcats of the weather world. They more than make up in fury what they lack in size. It has been estimated (because measuring instruments are always smashed) that these storms generate air currents which may attain a velocity of 300 mph, the highest air speed produced by nature. They charge in with the deep roar of a large squadron of bombers and can be heard 25 miles distant. In 1931 a particularly fierce tornado picked up an 83-ton railroad coach and its 117 passengers and whirled it 80 feet through the air, before dumping it into a Minnesota ditch.

Tornadoes usually form in the heat of the afternoon, several thousand feet above the earth's surface under certain atmospheric condi-

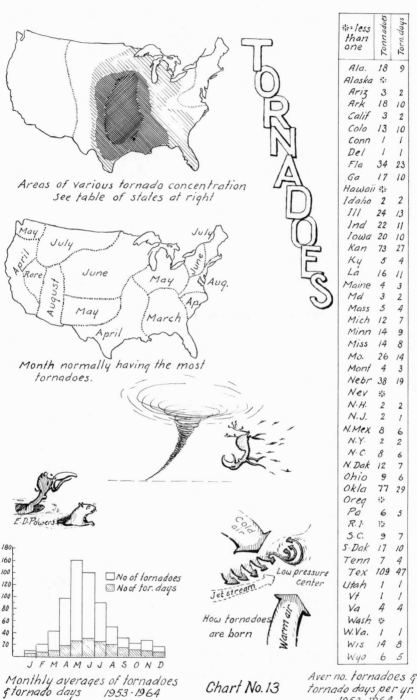

Areas of various tornado concentration
See table of states at right

Month normally having the most
tornadoes.

E. D. Powers

Monthly averages of tornadoes
& tornado days   1953-1964

Chart No. 13

Aver. no. tornadoes &
tornado days per yr.
1953-1964

✳ = less than one	Tornadoes	Torn. days
Ala.	18	9
Alaska	✳	
Ariz	3	2
Ark	18	10
Calif	3	2
Colo	13	10
Conn	1	1
Del	1	1
Fla	34	23
Ga	17	10
Hawaii	✳	
Idaho	2	2
Ill	24	13
Ind	22	11
Iowa	20	10
Kan	73	27
Ky	5	4
La	16	11
Maine	4	3
Md	3	2
Mass	5	4
Mich	12	7
Minn	14	9
Miss	14	8
Mo.	26	14
Mont	4	3
Nebr	38	19
Nev	✳	
N.H.	2	2
N.J.	2	1
N.Mex	8	6
N.Y.	2	2
N.C.	8	6
N.Dak	12	7
Ohio	9	6
Okla	77	29
Oreg	✳	
Pa	6	5
R.I.	✳	
S.C.	9	7
S.Dak	17	10
Tenn	7	4
Tex	109	47
Utah	1	1
Vt	1	1
Va	4	4
Wash	✳	
W.Va.	1	1
Wis	14	8
Wyo	6	5

TORNADOES

Cold air

Jet stream

Low pressure center

How tornadoes are born

Warm air

No of tornadoes
No of tor. days

tions and in association with heavy rains and often hail and thunder-storms. Eighty percent occur between noon and midnight. While they may appear any place in the world, and none of our states is completely free of them, no area is more favorable for their formation than the south and central plains of the United States.

By far the highest frequency is during spring and early summer, but the center of concentration migrates throughout the peak season. Starting in February and March over the central Gulf states, it spreads northward, southward, eastward and westward in April as indicated on the map. By May it has traveled to the southern plain states and in June is active in the Great Lakes areas, the northern plains and as far east as western New York state. The reason for this movement is the increasing penetration of warm moist air while contrasting cool, dry air still surges in from the north and northwest. It is here that tornadoes are born.

The whirlwind creates a partial vacuum which, being the opposite force to crushing pressure, can cause buildings caught in its destructive path to explode outwardly. Tornadoes usually move from southwest to northeast, at the rate of about 25 to 40 mph. The average length of travel is 10 to 40 miles but can exceed 300 miles. One may skip along touching down at points several miles apart. Lashing the earth with a roar, destruction is almost instantaneous and often complete. The damage is quite local and the sharply cut path is usually about 250 yards wide.

In addition to their violence, the hazard from hurricanes is heightened by their often sudden appearance and the erratic path of travel. The very efficient Environmental Science Services Administration, together with its rather recent National Severe Storm Forecast Center at Kansas City, not only broadcasts warnings of actual tornadoes but studies and observes closely areas in which there exist the particular conditions for producing these storms.

While property damage is very great, more important are the great number of deaths. Back in 1925, in a series of eight tornadoes throughout the Midwest, 740 people were killed in a single day. More recently, on April 11, 1965, again in the Midwest, 271 persons died in tornadoes.

As foreboding as this may all sound, people live, work and play in this vast area much as in the remainder of the land, but when driving your car during the active season and in areas of the greatest frequency, it's well to listen to the local weather broadcasts.

HAY FEVER

See Chart No. 14

To those unfortunates who are allergic to airborne pollens, the hay fever season can be a time of great misery.

As indicated on the Hay Fever Map, the entire eastern half of the country, particularly the upper portion, is heavily infested with various types of ragweed. A remarkable exception is most of Maine and New Hampshire which, except for a few counties in each, are rated from good to excellent. Aside from those two states there are few large areas in this whole vast region where sufferers can find relief.*

The ragweed season in most areas is confined to August and September, but along the Gulf Coast the weeds are still active in October. Southeastern California and Arizona have an early spring ragweed season (March, April and early May), which is more important locally than the fall occurrence. In central Florida and the vicinity of Tampa on the west coast the season begins in June and lasts until November. The same is true in the southern tip of Texas—around the Brownsville area. The duration and degree of air pollution can, however, vary somewhat from year to year because of differing weather conditions.

The following is the explanation of the designations used on the Hay Fever Map and in the text:

Excellent—This category includes places with perfect or near perfect ragweed pollen records.

Good—Indicates somewhat more pollen contamination but unlikely that anyone will experience severe hay fever symptoms.

Fairly good—A good degree of relief during most of the season.

Not recommended—The larger rings on the Hay Fever Map indicate heavy concentrations and the smaller rings lesser amounts.

There are about a hundred species of ragweed. The sagebrush is not in this category but is closely related antigenetically and the seasons of dispersal of pollen, are about the same. The term *hay fever* however,

*The best directory listing the ragweed conditions throughout the United States is the excellent little leaflet "Hay Fever Holiday" compiled by Oren C. Durham for the American Academy of Allergy. It is published by and available from Abbott Laboratories, North Chicago, Illinois. The name Durham is synonymous with hay fever research and much of the best literature on the subject has been written by him. The Abbott Laboratories has gone far beyond commercial interests in developing information and relief for the many victims of hay fever.

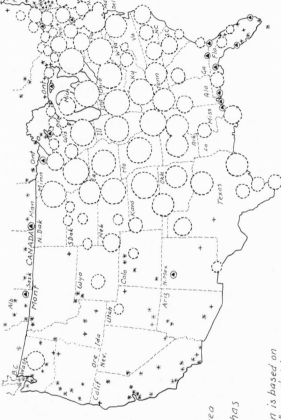

HAY FEVER

Chart No.14

Map symbols:
* ✳ Excellent
* + Good
* ◉ Fairly good
* ⌬ Disc size indicates pollen pollution

Ragweed season in most areas confined to Aug. and Sept. but: Gulf coast still active in Oct

S.E. Calif. & Arizona -- March, April & May can be more serious than the fall. Florida- central & Tampa area season - from June to Nov. Brownsville, Texas area has same season as Florida.

This sketch and information is based on the booklet, "Hayfever Holiday" compiled by Oren C. Durham, for the Pollen & Mold Committee of the American Academy of Allergy. It was published as a Public Service by Abbott Laboratories, North Chicago, Illinois.

includes any form of seasonal allergy which arises from the presence of airborne substance from plants.

ALABAMA. The Gulf coast at Foley, good; fairly good at Mobile. Field surveys throughout the remainder of the state reveal wide distribution of ragweeds in waste places and on farms. Birmingham has a very poor record.

ALASKA. No ragweed pollen was found as a result of atmospheric tests made in three places for one season, at Nome, Fairbanks, and Juneau.

ARIZONA. Excellent rating for the north and south rims of Grand Canyon. During the fall season in Phoenix conditions are excellent, but there is a spring ragweed season of at least moderate consequence. Our best information for the Tucson area gives it a rating of good for both spring and fall. For other communities in the state there are no atmospheric data.

ARKANSAS. The average exposure to ragweed pollen throughout the state is doubtless very heavy. No refuge areas are known.

CALIFORNIA. Excellent: Lassen Volcanic National Park, Sequoia National Park, Oakland, Sacramento, San Francisco, Monterey, Yountville, Yosemite National Park, Los Angeles, Pasadena, El Centro, Escondido, San Diego, Tujunga.
Good: Alpine, Arcata, Santa Barbara.
While air sampling has not been done in the great central valley, it is unlikely that any community there or elsewhere in the state will be found to have an appreciable degree of ragweed pollen pollution.

COLORADO. Excellent: Rocky Mountain National Park at Estes Park and Grand Lake, Mesa Verde National Park, Glenwood Springs, the crest of the Pikes Peak.
Good: Colorado Springs. Formerly this city's record was not so good. Ragweeds are not common on the west of the slope. Sagebrush is likely to be encountered in this area. Close exposure should be avoided by ragweed sensitive persons.
Denver and the east third of the state constitute an area of moderate to heavy ragweed exposure.

CONNECTICUT. Atmospheric studies have been made in 8 cities. No refuge areas are known.

DELAWARE. Field studies show ragweed to be abundant throughout. Nearest atmospheric studies are those made at Philadelphia and Baltimore. No refuge areas are known.

DISTRICT OF COLUMBIA and adjacent areas of Maryland show heavy ragweed pollen incidence.

FLORIDA. Excellent: Santa Rosa Island, Key West, Fort Myers, Miami Beach, Coral Gables, Miami, Sunnyside Beach (Panama City).

Good: Daytona Beach, Orlando, Sebring, Bradenton, Everglades National Park, St. Petersburg, Fort Pierce, Live Oak, West Palm Beach.

Fairly good: Fort Lauderdale (Beach), Jacksonville, Tallahassee, Tampa, Clearwater, Pensacola.

Not Recommended: Ocala, Gainsville, Melbourne, Panama City. The beaches of Florida are almost uniformly desirable; inland areas often not so good.

GEORGIA. Valdosta (only one season), fairly good, central and northern Georgia and the coastal area, as judged by tests at Atlanta and St. Simons Island and as checked by widely scattered field surveys, have moderately heavy exposure.

HAWAII. No significant amounts of any kind of ragweed have been found anywhere on the larger islands except in the area between Scoffield Barracks and Pearl Harbor on Oahu. Honolulu is probably ragweed-pollen free on account of prevailing northeast tradewinds. No daily atmospheric tests have ever been recorded.

IDAHO. Excellent: Sun Valley (2 years), Moscow.

Good: Boise, Pocatello.

All mountainous areas are likely excellent, but exposure to sagebrush pollen is possible throughout most of the state. Close contact with sagebrush should be strictly avoided.

29

ILLINOIS. No refuge area. Heavy records in 17 cities and towns.

INDIANA. No refuge area. Heavy records in 7 cities and towns.

IOWA. No refuge area. Heavy records in 6 cities and towns.

KANSAS. No refuge area. Atmospheric ragweed pollen incidence diminishes westward.

KENTUCKY. No refuge areas are known, but are barely possible in the Cumberland Mountains.

LOUISIANA. Heavy atmospheric pollution at New Orleans. Air sampling has been carried on only at New Orleans and at Vicksburg, Mississippi, across the river from Tallulah, Louisiana.

MAINE. Excellent: St. Francis, Greenville Junction, Millinocket, Presque Isle, Macwahoc, Quoddy Head, New Portland, Newagen, Enfield, Deblois, Belfast, Allagash, Grand Lake Stream, Bethel, Eagle Lake, Lincoln, Oquossoc, Speckle Mountain, Upper Dam.
   Good: Houlton, Newport, Jackman, Machias, Bar Harbor, Boothbay Harbor.
   Fairly good: Eastport, Rockland, Southport, York, Augusta, Camden, Rangeley, North Augusta, Orono.
   Not Recommended: Stonington, Poland Spring, Auburn, Alfred, Portland, Kineo.

MARYLAND. No refuge area is known. No atmospheric studies have been made in the mountainous parts of western Maryland.

MASSACHUSETTS. Good: Annisquam, East Gloucester, West Gloucester, Magnolia, Rockport, Nantucket Island.
   Fairly good: Gloucester, Worcester.
   Not recommended: Winchester, Boston, Northampton, Amherst, Newton Center.
   This is the state, and Boston the chief city, from which ragweed hay-fever victims first fled to the mountains and rocky coastal areas of New Hampshire and Maine some 100 years ago. Even so, ragweed pollen is much less abundant in the air in Boston than in most of the

larger cities of northeastern United States. Of the fourteen communities tested none offers excellent refuge conditions. Neither the Berkshires nor Cape Cod has received attention. Weed destruction seems to be effective on Nantucket Island. Otherwise ragweeds take over all waste areas.

MICHIGAN. Excellent: Isle Royale National Park.
Good: Sault Ste. Marie, Copper Harbor.
Fairly good: Houghton.
Fifty years ago much of the area of northern Michigan was doubtless entirely free from ragweeds and ragweed pollen, but sampling done in 57 systematically selected communities during the past 25 years has shown that no effective refuges remain in the lower peninsula, and that those of the northern peninsula are few, as listed above. The following list does not include any city of the lower peninsula. Those toward the beginning of the list are much better than those toward the end and might be suitable for persons with moderate sensitivity.
Not recommended (upper peninsula only): Saint Ignace, Blaney, Munising, Ironwood, Mackinac Island, Newberry, Powers, Menominee, Escanaba.

MINNESOTA. Fairly good: Tower, Virginia. Other places as good as or better than Tower and Virginia could likely be found in other parts of the Arrowhead County (northeastern corner of the state).
Not recommended: Duluth, Rochester, Minneapolis, Winona, Moorhead. The state has been inadequately covered.

MISSISSIPPI. Biloxi, on the coast, is fairly good. Field studies reveal an abundance of ragweed on farms throughout, so except for the immediate coast no refuge areas are likely to be found.

MISSOURI. No refuge areas.

MONTANA. Excellent: Glacier National Park at Belton and Many Glacier, West Yellowstone. Judging by the excellent records for more than 20 cities and towns in the adjacent parts of Alberta and Saskatchewan and at Yellowstone National Park, most of Montana is practically free of ragweeds.
Good: Miles City.

Very meager data, and no recent studies for this state. Sagebrush is widely distributed and should be avoided by persons known to be ragweed sensitive.

NEBRASKA. No refuge areas, but considerably less ragweed is found in the western third of the state than in the eastern part.

NEVADA. Very meager data, and no recent air sampling. Ragweeds are rare along the principal highways. Reno, excellent. Lake Mead, excellent in the fall, good in the spring ragweed season. Sagebrush is a possible factor.

NEW HAMPSHIRE. Excellent: Moosilaukee, Pawtuckaway, Errol, Lancaster, Carrol, Laconia, Colebrook, Blue Job Mountain, Derby, Groveton, Lincoln, Pittsburg, Warren, Whitefield.

Good: Bath, Conway, Dixville, Littleton, North Conway, Ossipee, Hampton, Plymouth, Bethlehem, Crotched Mountain, Dover, Franklin, Hillsboro, Holderness, New London.

Fairly good: Claremont, Concord, Federal Hill, Keene, Berlin, New Ipswich, Manchester, Weirs.

Not recommended: Hinsdale, Charlestown, Rye, Rochester, Lebanon, Jeremy, Exeter, Peterborough, Nashua.

NEW JERSEY. No refuge areas are known. Those places along the northern shore where relief is sometimes found are subject to high counts when the wind blows from the west. Studies have been made in 29 cities.

NEW MEXICO. Very meager atmospheric data. Ragweeds are probably comparatively rare throughout the state. Roswell is good, and Albuquerque fairly good.

NEW YORK. The reports on Long Island have produced variable records. Fire Island at Ocean Beach is sometimes fairly good, and Montauk likewise fairly good. No other records are available for the Island, except in Brooklyn which is not recommended.

Adirondack area. Excellent: Keene Valley.

Good: Blue Mountain Lake, Elk Lake, Keene, Loon Lake.

Fairly good: Big Moose, Chilson, Indian Lake, Long Lake, McColloms, Raquette Lake, Paul Smiths, Redford, Wanakena, Chateaugay Lake, Inlet, Sabattis, Schroon Lake (Severance), Tupper Lake, Newcomb, Owl's Head, Lake Placid, Mc Keever.

Catskill area. Good: Big Indian, Haines Falls, Pine Hill.

Fairly good: Fleischmanns.

Studies have been made in 85 other communities, including all of the larger cities, none of which can be recommended.

NORTH CAROLINA. No refuge areas are known, but air tests at Newfound Gap, Tennessee, on the crest of the Great Smoky Mountains prove the immediate area to be good. Likely there are other places equally good at similar or higher elevations in North Carolina. There are records of heavy concentration of ragweed pollen for four of the large cities of the state.

NORTH DAKOTA. No atmospheric data are available except in the narrow Red River Valley at Fargo (see map). No refuge areas are known, but conditions are likely much the same in the southern half of the state as in South Dakota. Judging from data from adjacent areas in Canada, there might be some good places along the northern edge of the state.

OHIO. No refuge areas. Adequate sampling has been done in seven large cities.

OKLAHOMA. No refuge areas. Adequate sampling has been done in seven large cities.

OREGON. No atmospheric studies have been made in eastern Oregon except at Milton-Freewater which is good.

Excellent: Coquille, Corvallis, Eugene. Crater Lake National Park, Portland, Turner.

PENNSYLVANIA. No refuge areas are known. Claims for mountain resorts have never been proved. Sampling has been carried on in ten large cities for many years.

RHODE ISLAND. No refuge areas.

SOUTH CAROLINA. No refuge areas are known, but our data are very meager. Nothing recent.

SOUTH DAKOTA. There are no refuge areas better than fairly good.
Fairly good: Rapid City, Mobridge.

TENNESSEE. No refuge areas are known. Along the crest of the Great
Smoky Mountains at Newfound Gap conditions were found to be good.
There are no accommodations at this point, but there might be places
with similar conditions at similar or higher elevations.

TEXAS. Big Spring is the only community out of the ten where studies
have been made which has a rating of good. Most of Texas is badly
infested with ragweeds. However, they diminish considerably toward
the west corner of the state, for example in El Paso.

UTAH. Excellent: Zion National Park, Bryce Canyon National Park.
Fairly good: Vernal in the extreme northeast corner of the state and
Hurricane in the extreme southwest corner of the state. The average for
metropolitan Salt Lake City is fairly good, except for the Canyon Rim
area.

VERMONT. Very meager data. Conditions on the east side of the state
are probably comparable to adjacent areas of New Hampshire. Heavy
atmospheric contamination is found in the upper Lake Champlain area.

VIRGINIA. No excellent or good refuges are known.

WASHINGTON. Excellent: Seattle-Tacoma Airport, Mt. Rainier National Park (Longmire, White River, Paradise Valley), Seattle, Olympic
National Park, Spokane, Yakima.
Good: Walla Walla
Except for the badly ragweed-contaminated orchards in the immediate vicinity of Wenatchee, all but one place among the ten tested in the
state are excellent or good. The large disc on the map of Washington
represents conditions found only at Wenatchee.

WEST VIRGINIA. No refuge areas are known.

WISCONSIN. No refuge areas are known, but no adequate investigation
has been made in the vast lake region of the northern part of the
state.

WYOMING. Very meager data except at the national parks.
Excellent: Grand Teton National Park, Yellowstone National Park.
Lander is not recommended.*

SMOG AND AIR POLLUTION

See Chart No. 15

Ecology may be a rather new word to most of us but under other
names it has long been the concern of conservationists and those unfortunates troubled with respiratory problems. Thankfully, we are all now
awakening to the dangers, but the necessary major improvements will
not happen overnight. In the meantime we can only search for those
locations where the air is pure or at least, less polluted.

Strange as it may seem, such spots are not all that easy to identify.
There is an almost complete lack of agreement as to the type and degree
of concentration which may be injurious. As an example, the list of the
20 most air-polluted cities shown on the map will not coincide with the
many selections that you will see published. Another such list before
me, agrees with the first four cities but rates the remaining 16 as follows;
5. Cleveland; 6. Pittsburgh; 7. Boston; 8. Newark; 9. Detroit; 10. St.
Louis; 11. Gary–Hammond–East Chicago; 12. Akron; 13. Baltimore;
14. Indianapolis; 15. Wilmington; 16. Louisville; 17. Jersey City; 18.
Washington; 19. Cincinnati; 20. Milwaukee. There seems to be as much
divergence of opinion here as in the ratings of the ten best dressed
women.

There is ample reason for this confusion, since there is no long
statistical history of this subject on record. Indeed there are no generally accepted standards of the types and concentrations of pollutants
which may be injurious. Adding to that are the great fluctuations
possible in actual conditions, within a year or even a day, which illus-
trate just how complex the problem is.

There seems to be some acceptance that at least the major contaminants appear to be carbon monoxide, oxidants, nitrogen dioxide, sulphur dioxides, hydrocarbons and airborne solid particles. Also that the
chief offenders are automobiles, industry, electric power plants, space

*The above information regarding ragweed was abstracted from the excellent booklet, "Hay
Fever Holiday," which was compiled by Oren C. Durham for the Pollen and Mold Committee
of the American Academy of Allergy.

Air pollution rating of cities (No 1, highest pollution)

1 New York City
2 Chicago
3 Philadelphia
4 Los Angeles
5 Pittsburgh
6 Newark
7 Detroit
8 Buffalo
9 Louisville
10 Washington
11 Paterson
12 Denver
13 Niagara Falls
14 Baltimore
15 San Francisco
16 New Haven
17 Portland
18 Dallas
19 New Orleans
20 Hartford

Air pollution in the cities.
There are many such lists
with various ratings of
cities.

SMOG
& AIR
POLLUTION

Chart No. 15

heating and refuse disposals. The following table shows some approximations.

SOURCES OF AIR POLLUTANTS							
(Millions of tons per year)							
Sources	Millions of tons per year	Percent of total	Carbon Monoxide	Sulfur Oxides	Hydro-carbons	Nitrogen Oxides	Particles
Autos	86	60%	66	1	12	6	1
Industry	23	17%	2	9	4	2	6
Electric Power Plants	20	14%	1	12	1	3	3
Space Heating	8	6%	2	3	1	1	1
Refuse Disposal	5	3%	1	1	1	1	1
Total	142	100%	72	26	19	13	12

To better appreciate just how many factors may be involved in a single situation, let's consider an industrial plant (all such operations are said to produce only 15 to 20% of the total air pollution). Under normal operation, its emissions may be acceptable; but if production is raised to over designed capacity, or there are equipment or other operating troubles, the pollutants may increase drastically. Then again, on a day with a brisk breeze, the concentration may be diluted to very low limits. Conversely during an air inversion, even with good operation, the air may quickly reach very dangerous levels of pollution. The weather condition can also be a most important factor in regard to the behavior of the most serious offender, the automobile. The infamous "photo-chemical smog" of Los Angeles is caused by the action of sunlight on the contaminants in the exhaust fumes, which converts some of them into additional poisons. If there are strong updrafts of air, the fumes are quickly diluted and dissipated in the general air mass. The problem develops with an air inversion: a layer of cool air capped with a warmer one, which blankets it to the earth, entrapping the fumes.

All of this is of interest and a bit frightening, but what can we do about it? Unfortunately there is little literature, or organized advice available, and this will of necessity have to be a "find it yourself" undertaking.

The logical first reaction would be to stay away from the most serious offenders. While the "twenty worst cities" lists don't agree, any place included on one must be suspect. It's well to remember, however, that a "bad rating" doesn't necessarily indicate the year-round conditions. The source of the pollution may operate on a seasonal basis. Prevailing winds or other weather conditions may also limit the period. Los Angeles, for instance, has a seven-month smog season.

The worst offender of all—the automobile—is not so easy to avoid, nor are its poisonous fumes so readily detected. Areas with high traffic density will be polluted, but here again weather can affect the local portions. As an example, the prevailing wind along the east coast of Florida is off the ocean, and people living close to the seashore enjoy clean, pure air. Those only a relatively short distance inland, but to the west of the very heavily trafficked highways and more dense population centers, may be subjected to injurious concentrations of fume pollution.

Fuel-burning electric power plants and industrial operations are more obvious, but they don't all cause the same degree of hazard. Sometimes local inquiry will produce useful information, but again it will be well to consider the weather and prevailing winds. Observing the condition of the surrounding vegetation, and even the paint on houses, will sometimes identify a particularly bad area. In the southeast and along the Gulf it's well to inquire if there is a paper pulp mill in the vicinity. You may be in a delightful little town for several days before the wind shifts and you are driven out. Because many of these good and bad areas can be so localized, and with the many variables and changeable factors, it will be a long time before a comprehensive directory can be produced.

2

The Forty-Eight States

See Chart No. 16

This is the longest but hopefully the most useful portion of the entire book. You will note that we have treated the country as twelve climatic areas rather than as fifty separate states. It is arranged so that when planning a trip, the reader can first refer to the overall maps (chapter 1) and then study in greater detail the various sections through which he may travel or stay.

To obtain the maximum value, the reader should compare the data concerning the cities and places being considered with the tabulated figures of his own part of the country. For most people, such statistics as 42 inches of rainfall, or even relative humidity figures in themselves, have little real meaning. If compared with the familiar conditions in one's home area, however, they should have significant value in contrast.

So, too, it is not always very meaningful to describe a place as being pleasant or comfortable unless related to a particular individual's taste. A person accustomed to the crisp Minnesota country could die a thousand deaths during the humid months in Mobile, whereas to one from the northern part of Alabama, it might spell relief and comfort. While neither Florida nor the French Riviera may be a tropical paradise in January, they are much more pleasant than Chicago or Paris at that time.

The tabulated figures in particular can be useful from another viewpoint. The goal of the vacationist is perfect weather during a specific limited period. The retiree is looking for overall good conditions or at least the shortest possible intervals of really uncomfortable weather, be

39

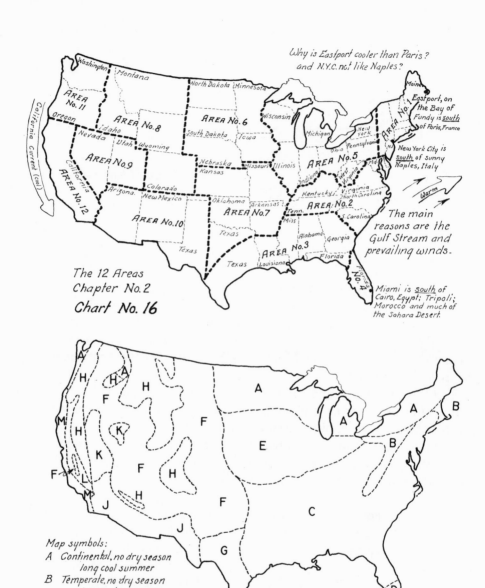

Why is Eastport cooler than Paris?
and N.Y.C. not like Naples?

California Current (cool)

AREA No.11

AREA No.8

AREA No.6

AREA No.5

AREA No.1

Washington
Montana
North Dakota
Minnesota
Oregon
Idaho
South Dakota
Iowa
Wisconsin
Michigan
Maine
New York
Ohio
Pennsylvania
Indiana
Kentucky
Virginia
North Carolina
West Virginia

AREA No.9
Nevada
Utah
Wyoming
Nebraska
Kansas

AREA No.12
California
Arizona
Colorado
New Mexico

AREA No.10
Oklahoma
Arkansas
Tenn.
Miss.
Texas
Louisiana

AREA No.7

AREA No.2
S. Carolina

AREA No.3
Alabama
Georgia
Florida

AREA No.4

Eastport, on
the Bay of
Fundy is south
of Paris, France

New York City is
south of sunny
Naples, Italy

Warm

The main
reasons are the
Gulf Stream and
prevailing winds.

Miami is south of
Cairo, Egypt; Tripoli;
Morocco and much of
the Sahara Desert.

The 12 Areas
Chapter No. 2
Chart No. 16

Map symbols:
A Continental, no dry season
     long cool summer
B Temperate, no dry season
     long cool summer
C Subtropical no dry season, hot summer
D Tropical savanna rainy - dry winter
E Continental no dry season - hot summer
F Middle latitude steppe, aver. annual temp.
     below 64°
G Subtropical steppe, aver. temp. over 64°
H Highlands, various local climates
J Subtropical desert, aver annual temp.
     above 64°
K Middle latitude desert, aver annual
     temp below 64°

AREAS & CLIMATES

L Subtropical dry hot summer
M Mediterranean, long cool dry
     summer.

it temperature, humidity, winds or whatever his particular aversion may be.

In studying the tables, covering weather in individual cities, the reader will note some places where the undesirable conditions are concentrated within a few months. Others may have two or several such spells or perhaps only one extended over a much longer portion of the year. A spot like Gunnison, Colorado, can be a summer vacationist's dream but hardly ideal as a retirement haven unless one can go into hibernation before the snow drifts reach the eaves.

There are, however, a few fortunate spots like San Diego, California, which could be the answer for both. With almost ideal year-round climate, it attracts Canadians and others from the northern climates in winter, while the refugees from Phoenix, Arizona, and such places find welcome relief there during July and August.

The following pictorial maps will probably be readily understood but the tables describing the weather conditions throughout the year in some 100 cities may at first glance appear as formidable as a Chinese timetable. Don't be frightened off by this array of figures. We have translated the climatologist's hieroglyphics into much simpler language.

You may wonder why there are so few figures in the Temperature Humidity Index column. Blank spaces indicate "no humidity problems." T.H.I. is a combination of temperature, humidity, and a few other factors. 72 is the magic number to remember. Below that, most people are comfortable but higher figures can spell unhappiness.

When the T.H.I. is below:

72—most people think they're comfortable;
75 and above—many know they're not, and head for the swimming pool or air-conditioning;
80 and over—if they can afford it, most start packing their bags.

If this seems like an oversimplification and anyone really wants to know why he is uncomfortable, we refer him to the T.H.I. conversion chart and detailed explanation in chapter 7.

Another relatively new but very important comfort guage that has recently come into general use is the "Wind Chill Factor." This also is simple. Have you ever been surprised at how low the mercury actually registered on a day that didn't feel particularly cold? This is most

41

noticeable in an atmosphere of still air or in a protected location. And conversely muttered curses through chattering teeth when the weatherman, from his comfortable studio, cheerfully announced "and it's only 36 degrees this fine morning." In the latter case, wind was probably the culprit, causing the chills and misery, even though the mercury was not really very low.

As an example, the Boston chart shows a minimum average temperature of 27° in December. That may not appear too frigid but, because of the chilling effect of the wind, the "Wind Chill Factor" scale will read 4°! In other words, when the thermometer tells you it's 27°, your bones know it feels like 4°. The Air Force discovered just how important this measurement really is. There is a conversion table chart and a more detailed explanation of "Wind Chill Factor" in chapter 7.

Also on these tables, there is a column labeled "Storm Intensity." The symbols indicate the type of storms which may be expected during the particular month and how severe they are apt to be. The first symbol indicates the predominating type of storms expected each month; the second symbol, the severity of the storms.

Symbols and Type of Storm

> R — rainstorm
> S — snowstorm
> T — thunderstorm

Intensity of Storm

> 1 — light
> 2 — light to moderate
> 3 — moderate
> 4 — moderate to severe
> 5 — severe

For example, a severe thunderstorm would be shown as T5.
Also, the sequence of symbols indicates the type of storm.
RS means rain more likely than snow.
SR means snow more likely than rain.
The other columns in these tables are probably reasonably understandable but such measurements as "inches of rain" may be a bit difficult to actually visualize. Again we suggest that you compare these

figures with those shown for your hometown or other areas with which you may be familiar.

The column "Precipitation in inches" is the standard designation of the U.S. Weather Bureau. It is shown in two vertical rows of figures. The heading "Total" means the sum of all forms of precipitation, rain, snow, hail, etc.; the second row is total inches of snow only. As an approximation, we say that ten inches of snow equal one inch of rain (actually this varies, depending upon how fluffy or damp the snow may be). For all practical purposes "precipitation" is made up of rain and snow. Therefore if we convert the inches of snow into inches of rain (dividing inches of snow by ten, equals inches of rain) and by subtracting that figure from the total inches of precipitation we get the approximate inches of rain.

We agree that this seems a bit cumbersome at first, but you will find by trial that it's quite simple. Again when indicating the frequency of precipitation, we show the average number of days per month and year that there will not be even 0.01 inches of precipitation. As you can well imagine, 0.01 inch of moisture isn't very much—often hardly a good sprinkle. You might wonder why a higher figure isn't used—perhaps at least ⅛ inch of rainfall. That would certainly seem more practical and useful. The explanation is that most of the figures are backed by statistics which extend back forty or fifty years. From the beginning, all precipitation was measured and recorded as a total figure on the basis of 0.01 inch or more. Until quite recently, weather information was oriented almost exclusively to the requirements of the professional meteorologist, and these were the only continuously recorded figures available.

Some of the area descriptions may seem to rather over-emphasize unfavorable conditions. Actually, of course, if there were no clouds, storms, etc., the sun would be visible the entire period of daylight hours. To a great extent, we are relating the frequency of these occurrences and the portion of the time the sun is partly or fully obscured.

In each area description, we have included a table of the percent of possible sunshine, on a monthly and annual basis, in a number of key locations throughout that particular section. There is a wide range from the low average 18% in Roseburg, Oregon during December to the amazing June 98% in Yuma, Arizona. The detailed maps of the areas also show the total hours of sunshine per year in most of these places.

43

There are also several important maps in chapter 1, such as the one showing the average number of hours sunshine per day in summer and winter across the United States. This varies from a scant 2 or 3 hours per day of mid-winter sunshine in the northwest to a summer high of 12 hours per day in the southwest. There are also maps showing the summer, winter and annual averages of percent maximum possible sunshine in these forty-eight states.* Additionally, there is the map covering the number of clear days which ranges from 80 to 260 days per year. In contrast, we include the average number of cloudy and foggy days per year. Availability of sunshine is of prime importance to traveler, vacationist and retiree alike, but for the full story, the sun must be considered in conjunction with all other elements of weather.

THE NORTHEAST—Area Number 1

See Charts Nos. 17 and 18

(Maine, New Hampshire, Vermont, Massachusetts, Rhode Island, Connecticut, Eastern New York, Eastern Pennsylvania, New Jersey)

This interesting area has long been the year-round playground for many. Enjoying an unusually wide range of seasons, it includes just about every type of scenic attraction imaginable. It is also becoming increasingly popular with retirees who like the quarterly change which is typical of a continental climate.

While it is true that each segment of the year includes bad weather, none lasts too long and can easily be avoided by switching to indoor activities. Many think of the northeast as a hectic industrial complex, enveloped in smoke, fumes and Wall Street ticker tape. The newcomer may be pleasantly surprised to discover that the whole northwestern stretch of this entire area is "green country,"—forests, woodlands, rolling meadows, hilly lands and mountains, with many lakes and streams. There are, indeed, many industrial concentrations but little reason for either the vacationist or retiree to tarry very long in any of them.

Like most of the upper eastern third of the United States, this area is classified as having a humid continental climate. That means a great

*Alaska and Hawaii are treated separately.

variety of weather conditions and sudden changes. Winters are usually cold with sharp, snappy waves and even an occasional blizzard always is a possibility. The interior, particularly at the higher elevations, is most often cold, snowy, dry and great for outdoor winter activities. The coastal areas, suffer from much wet, windy, miserable weather—a perfect place for indoor hibernation.

Summers will generally be hot with frequent short thundershowers and much sunshine. Spring can be a delight. When fortune smiles this whole area can rival the famed English springtime countryside. But all too often there is no spring at all. Instead there can be many days of biting winds, chilling rains and mud up to here. Autumn is more dependable. Clear, crisp sunny days are almost a certainty, and the foliage color change is the most spectacular in the world.

Maine, New Hampshire, Vermont, and Northeastern New York

WINTER (interior)

The continuous snow cover during this season is an invitation to all outdoor winter activities. The principal ski areas, which are in New Hampshire, Vermont and northeastern New York (Adirondacks), usually average 80 to 100 inches of snowfall per winter. January and February is the time of heaviest snow, with approximately 19 inches per month in the Burlington, Vermont area. The snows fall both from heavy coastal storms which dump about 10 to 15 inches at a time and the less intense storms from the west which move eastward across northern New York State into the New England States, dropping 2–4 inches per storm. These lighter snowfalls occur at least once a week. The winter precipitation pattern in Maine is much the same. Winter skies are mostly cloudy. Clear sunny days are rare, one sunny day in seven being a good average. Light snow flurries are common.

The daytime temperatures in January and February in the mountains usually climb into the upper twenties but do not rise above freezing. At night the mercury falls to about 8 or 9°. December is less cold with daytime temperatures in the low thirties and daily minimums in the teens. Two or three times a winter this area is hit by a severe cold wave which drops temperatures to −30° and on occasion, −40° has been recorded. Because the very low temperatures occur in the early morn-

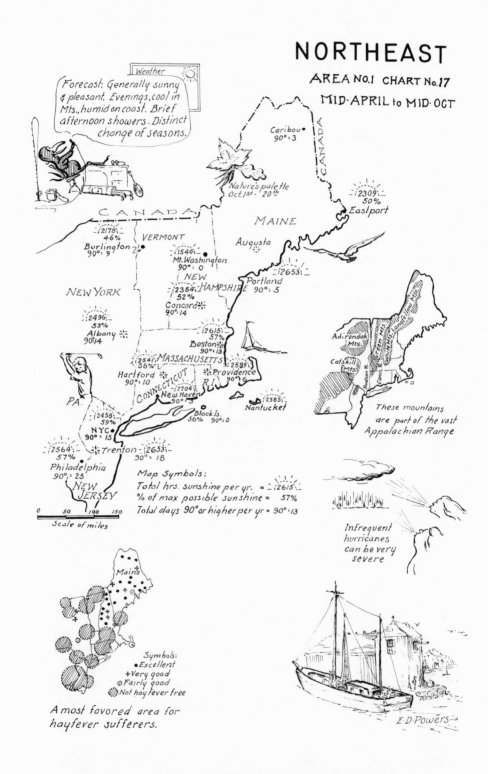

NORTHEAST

AREA No.1 CHART No.17

MID·APRIL to MID·OCT

Weather

Forecast: Generally sunny & pleasant. Evenings, cool in Mts., humid on coast. Brief afternoon showers. Distinct change of seasons.

Caribou 90°=3

Nature's palette Oct. 1st - 20th

CANADA

MAINE

2309 50% Eastport

2178 46% Burlington 90°=9

VERMONT

1540 Mt. Washington 90°= 0

NEW HAMPSHIRE

Augusta

2653 Portland 90°=5

NEW YORK

2354 52% Concord 90°=14

2496 53% Albany 90°=14

2615 57% Boston 90°=13

MASSACHUSETTS

2541 56% Hartford 90°=10

2589 Providence 90°=6

R.I.

CONNECTICUT

2704 New Haven 90°=

Nantucket

Adirondak Mts.

Green Mts.

White Mts.

Longfellow Mts.

Hudson R.

Conn. R.

Catskill Mts.

These mountains are part of the vast Appalachian Range

PA.

2458 59% NYC 90°=15

2583 56% 90°=0 Block Is.

2564 57% Philadelphia 90°=25

2653 Trenton 90°=18

NEW JERSEY

0   50   100   150
Scale of miles

Map Symbols:
Total hrs. sunshine per yr. = 2615
% of max possible sunshine = 57%
Total days 90° or higher per yr = 90°=13

Infrequent hurricanes can be very severe

Maine

Symbols:
• Excellent
+ Very good
⊕ Fairly good
▧ Not hay fever free

A most favored area for hayfever sufferers.

E·D·Powers

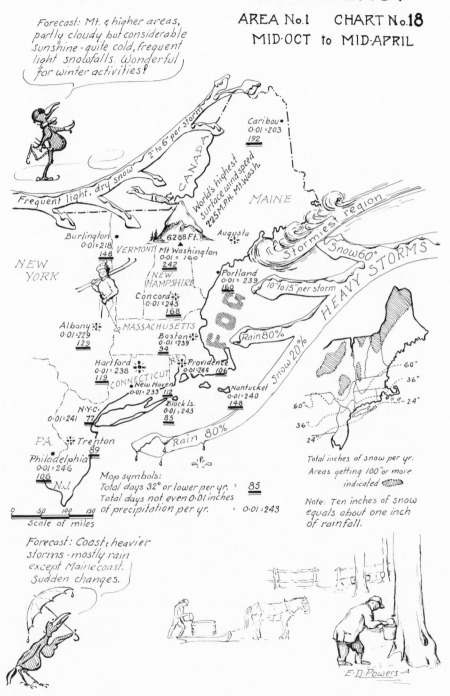

ing when the air is generally calm, the cold is not as biting as these figures might indicate. Moreover, the dryness of the air is also a factor which makes the cold more bearable.

WINTER *(coast)*

The coast of northern New England (Maine, New Hampshire) is slightly milder and wetter than the inland areas. The precipitation can be either in the form of rain or snow. This strip receives approximately 10 inches of precipitation of which more than one half is snow, approximately 60–70 inches. (Ten inches of snow equals about one inch of rain) Occasionally very severe east coast gales batter coastal areas with high winds and really heavy rain or snow.

Daily temperature highs in January and February are in the low thirties. Humidity tends to be higher than in the interior, and winds are slightly stronger causing these temperatures to feel somewhat more uncomfortable than the lower inland readings. At night temperatures generally drop into the low teens with the southern coastal sections in the mid to upper 30s, dropping into the low to mid 20s at night.

SPRING AND FALL

Seasons tend to lag; therefore, the summer warmth lingers into the autumn and winter wetness and chills extend into spring. Melting winter snows also make spring too sloppy underfoot for most outdoor activities. In both inland and coastal areas, the precipitation is slightly greater during the spring than fall. Spring follows the winter pattern, thus the coast receives more precipitation than inland sections. Rainfall in autumn is quite uniform throughout the northern New England region, averaging approximately ten inches. In Concord, New Hampshire one can expect rain (sometimes wet snow—during March, one out of three storms produces some snow) about three days a week during the spring. In the fall, however, that city averages rain only one to two days a week. In May thunderstorm activity is on the increase. Although usually not severe, and short in duration, they can drench an afternoon's activity.

In the fall there always exists the possibility of a hurricane which fails to follow its regular track. Two such hurricanes smashed through the northeast on September 21–22, 1938 and September 14–15, 1944. There were none then until ten years later, when late summer and early fall of 1954 and 1955 brought hurricanes to the northeast in numbers unmatched in Weather Bureau records. In 1954 there were three; one each in late August, in early September and mid October. In 1955 there was one in mid August and another less than a week later. In all of these storms, fierce winds and torrential rains combined to cause great devastation. In the interior, catastrophic river floods occurred. This period of such unusually high activity did not at all follow the historic pattern. As late fall approaches, surprise snowstorms can occur prematurely and precipitation in the form of snow can be expected. In the interior, snow will fall about three times out of ten. Interestingly enough, the effects of the ocean on coastal temperatures are insignificant in the spring and fall because the water temperature is about 45 degrees, which is approximately the same as the air. Thus there are little temperature differences between coastal and inland areas at this time. However, autumn temperatures are milder and more pleasant than in the spring. March winds and April showers, along with cloudy skies, can make spring less attractive.

SUMMER

In New England as far north as Maine, summers are warm and humid, with frequent heat waves and lots of sunshine. Normally T.H.I.'s during July and August reach an average of 76° when many people experience some discomfort. During particularly bad heat waves, T.H.I.'s can rise into the eighties, then almost everyone is uncomfortable. Along the coast, the effects of the ocean cause temperatures to be somewhat lower than inland, but the humidities are higher. Yet in comparing T.H.I.'s, those along the coast are still somewhat more favorable. In contrast to the hot days, the nights throughout the area are rather cool. Temperatures at night drop to the mid to upper fifties, providing delightful sleeping weather inland and slightly warmer in coastal sections.

It must be remembered that one can easily escape the heat in New England by heading toward the high country or the water. Fortunately, mountains and lakes are plentiful in this area and the sea coast is usually not too distant.

Sunshine is abundant through the summer interrupted only by brief afternoon thundershowers, mainly in the hilly sections. Prolonged rain is rare during this season.

Southeastern New York, Connecticut, Rhode Island, Long Island, Eastern Pennsylvania, New Jersey, and Massachusetts

WINTER

As a general rule, one in three days in this area experiences some form of precipitation, showery weather being most common; but the occasional heavier storm can dump large quantities. During the month of December, 20%–30% of the precipitation is snow and 70%–80% is rain in the inland sections. Along the coast the percentage of snow is even lower. The odds are against getting a severe snowstorm during early December, especially along the coast. Late December usually brings increased snowfall activity and the beginning of the ski season. January is generally a wetter month with several east coast storms developing. Contrary to belief, most of these storms drench the coast with rain rather than snow. Inland, however, half of these storms usually produce snow totaling up to ten or twelve inches. The Boston area is strongly influenced by the warm ocean waters and therefore less than 20% of the precipitation during the month of January is in the form of snow, totaling less than seven inches.

February is the "snow month" in this area. With ocean temperatures getting colder, even coastal areas are more threatened by snow storms. However in New York City and Boston and other immediate coastal sections rain still predominates over snow in seven of ten cases. Snowfall, near the ocean, during this month averages about nine inches with as little as 4–7″ in New Jersey. Inland areas experience ten to fifteen inches of snow. At least half of the precipitation in the interior sections falls as snow, making skiing conditions good to excellent in the mountains. Partly cloudy to cloudy days dominate the entire winter season.

Temperatures in this whole region vary widely depending upon location. In general, coastal sections are warmer and show smaller variations in temperature than inland. The mountains and elevated portions are substantially colder than coastal flatlands. In New York City, Boston, Nantucket, and the New Jersey shore, temperatures will be in the upper thirties to low forties during December, January and February.

At night they will usually dip into the low to mid twenties, with mid to upper twenties along the New Jersey coast. Readings below 10° are rather rare, especially at Nantucket. Maximum afternoon temperatures are usually in the low thirties, dropping to the low to mid teens at night. Occasionally cold waves will push the mercury below zero in these areas, sometimes registering twenty to thirty degrees below. Such extreme temperatures, while very rare, are possible even in the southern sections such as eastern Pennsylvania.

SPRING AND FALL

There is little difference in the amount of precipitation in spring and fall. Both seasons average approximately three inches per month. However, in the spring rainy days are comparatively more frequent than in the fall since the precipitation in the spring is in the form of light showers. Although March can still come under the influence of storm activity, April and May are particularly showery. In autumn the precipitation is heavier during the occasional east coast storms. As hinted above, snowstorms do occur in March and are sometimes the severest of the winter. Generally, however, the threat of snow diminishes rapidly as the month of March comes to a close. Winds, melting snows, frequent periods of precipitation combine to make March rather unpleasant. Snow in the beginning of November is not common but quite possible, while the chances of snow increase toward the end of the month.

The occurrences of clear sunny days are greater in the autumn, with the exception of November. Early spring is particularly characterized by quick changes. Often mild sunny days are followed by cold snowy ones.

Basically, coastal temperatures and inland temperatures during spring and fall vary only slightly. The reason being that it is somewhat cooler along the coast than inland during the spring and somewhat warmer during the fall. The difference, however, is insignificant.

Temperatures are very changeable during the month of March. Averages are quite deceiving. March could be below 10°F, especially inland, and then in a matter of days experience temperatures well up into the seventies or low eighties. Generally, however, temperatures reach the upper forties and low fifties on most afternoons and drop into the low thirties at night.

April shows a decided warming trend. Mid-fifties are common during most afternoons, reaching near 60° in southern sections and dropping into the mid to upper thirties at night. New Jersey will usually be in the low 40s. Coastal sections might average two or three degrees cooler. May is usually a very pleasant month. Temperatures are in the comfortable range, mid to upper sixties during the afternoon and dropping to the upper forties to low fifties at night. This is sweater weather during daytime hours and comfortable sleeping weather at night. Extremes can occur, but they are the exception.

Fall is somewhat milder than spring. September is still rather warm, especially inland. Daytime temperatures usually reach well into the seventies on most days. The occasional nineties are possible but uncommon during this month. At night one can usually expect that the thermometer will drop to the mid or upper fifties.

October temperatures are in the low to mid sixties during the afternoons, falling into the upper forties and low fifties at night. Northern inland areas experience low forties, with frost occurring several nights during the month.

November weather is rather chilly. Low to mid fifties are the rule during the afternoons, and low thirties and upper twenties in the inland areas at night. Along coastal sections and in southern portions of this area, temperatures remain in the mid to upper thirties at night. Frost is quite common during this month and occasional cold snaps can drop the thermometer below ten degrees. However, this is rare.

SUMMER

Summer months can sometimes be uncomfortable, due to the combination of high temperatures and high humidity. It will bother some people while others will be quite unaware of any discomfort. The latter part of July and early August are the hottest periods of the summer. In New York City and parts of New Jersey, July afternoon temperatures usually rise into the low to mid 80s and the T.H.I. reaches approximately 76. Immediate coastal sections usually remain in the upper 70s. These figures are the same for the beginning of August, dropping slightly toward the end of the month. Numerous heat waves elevate the mercury to the nineties and sometimes over 100° in New York City and the surrounding area. Many evenings are quite humid and only an air conditioner can provide relief from discomfort. June in

New York City is just slightly cooler with the T.H.I. approximately 75. During late May, early June, and the month of October the weather is usually very pleasant in New York City and Philadelphia. These are generally good periods for visits. But both should be avoided during the stifling July–August hot spells.

Connecticut, southeastern New York and eastern Pennsylvania have daytime temperatures that do not differ significantly from those in New York City. However, much of this area is hilly, especially Bear Mountain and the Catskills in New York, and the Poconos in Pennsylvania. Temperatures at higher elevations are somewhat cooler. Unlike New York City, the nighttime temperatures in this area (especially in the interior) are cool and comfortable. Average evening temperatures range from the upper fifties in the Catskills, for example, to the mid and lower sixties at Bear Mountain and in Connecticut.

Table 1	AVERAGE % OF MAXIMUM POSSIBLE SUNSHINE (% of daylight hours)												
Area Number 1 Northeast	Jan	Feb	March	April	May	June	July	Aug	Sept	Oct	Nov	Dec	Year
Maine													
Eastport	45	51	52	52	51	53	55	57	54	50	37	40	50
New Hampshire													
Concord	48	53	55	53	51	56	57	58	55	50	43	43	52
Vermont													
Burlington	34	43	48	47	53	59	62	59	51	43	25	24	46
Massachusetts													
Boston	47	56	57	56	59	62	64	63	61	58	48	48	57
Connecticut													
Hartford	46	55	56	54	57	60	62	60	57	55	46	46	56
Rhode Island													
Block Island	45	54	47	56	58	60	62	62	60	59	50	44	56
New York													
Albany	43	51	53	53	57	62	63	61	58	54	39	38	53
New York	49	56	57	59	62	65	66	64	64	61	53	50	59
New Jersey													
Atlantic City	51	57	58	59	62	65	67	66	65	54	58	52	60
Pennsylvania													
Philadelphia	45	56	57	58	61	62	64	61	62	61	53	49	57

PORTLAND, MAINE Elevation 61 Feet Table 2

Month	Temperatures Average Max.	Average Min.	Extreme Max.	Extreme Min.	T H I*	Wind Chill Factor*	Precipitation Total	Snow	Not even 0.01" precip.	More than ½" of snow	Clear	Cloudy	Thunder-storms	Fog	90° or higher	32° or lower	% of possible sunshine	Relative Humidity A.M.	P.M.	Wind M.P.H.	Direction	Storm Intensity*
Jan	31	11	64	−21		−7	4	20	19	5	16	15	0	2	0	30	55	81	65	9	N	R, S-3
Feb	32	11	64	−39		−7	4	17	17	5	17	11	0	2	0	27	62	81	61	10	N	R, S-3
March	41	22	86	−21		7	4	12	20	4	17	14	0	4	0	27	59	76	60	10	W	R, S-3
April	52	32	85	8		19	4	2	18	1	15	15	1	3	0	15	57	73	56	10	S	R, S-3
May	63	42	92	23			3	T	18	0	16	15	2	6	0	3	56	75	59	9	SSW	R-2
June	73	51	97	33			3	0	18	0	18	12	5	5	1	0	63	77	60	8	S	R-2
July	79	57	98	41	74		3	0	22	0	20	11	4	6	2	0	68	78	58	8	S	R-2
Aug	77	55	100	38	73		3	0	22	0	22	9	3	5	2	0	66	82	58	7	S	R-2
Sept	70	47	95	23			3	0	22	0	19	11	2	7	0	2	63	85	60	8	S	R-2
Oct	60	37	87	18		27	3	0	22	0	20	11	0	5	0	9	60	85	60	9	N	R-2
Nov	47	28	73	6		16	4	3	19	1	15	15	0	4	0	20	49	85	65	9	W	R, S-3
Dec	35	16	60	−15		1	4	11	21	3	18	13	0	2	0	28	54	82	62	9	WNW	R, S-3
Year	55	34	100	−39			42	67	238	19	213	152	17	51	5	161	60	80	60	9	S	

Notes:

T Indicates "trace"

* For full explanation of (T-H-I) "Temperature Humidity Index;" "Wind Chill Factor" and "Storm Intensity," see beginning of Chapter 2.

Average date of first freeze .... October 15
" " last " .... April 29
" freeze-free period .... 169 days
10 inches of snow equal approximately one inch of rain.

BURLINGTON, VERMONT — Elevation 331 Feet — Table 3

Month	Temperatures Average Max.	Average Min.	Extreme Max.	Extreme Min.	T H I *	Wind Chill Factor *	Precip. Total	Precip. Snow	Not even 0.01" precip.	More than ½" of snow	Clear	Cloudy	Thunder-storms	Fog	90° or higher	32° or lower	% of possible sunshine	Rel. Hum. A.M.	Rel. Hum. P.M.	Wind M.P.H.	Wind Direction	Storm Intensity *
Jan	28	8	63	−30		−12	2	14	17	5	11	20	0	1	0	29	34	81	69	11	S	S, R-3
Feb	28	8	60	−28		−12	2	15	15	5	11	17	0	1	0	27	44	79	65	11	S	S, R-3
March	39	20	84	−24		4	2	12	18	3	13	18	0	1	0	27	48	76	61	11	S	S, R-2
April	53	32	86	5		19	3	4	17	0	13	17	1	1	0	15	47	74	55	11	S	R, S-2
May	67	43	92	25			3	T	18	0	14	17	3	1	0	2	53	73	54	10	S	R-2
June	78	54	96	33	72		4	0	19	0	16	14	6	1	1	0	59	76	55	9	S	T-3
July	82	58	100	43	76		4	0	19	0	19	12	7	0	2	0	62	76	54	8	S	T-3
Aug	80	56	101	38	74		3	0	20	0	19	12	6	1	1	0	59	80	55	8	S	T-3
Sept	71	48	95	25			3	T	18	0	16	14	3	1	0	1	51	83	59	9	S	R-2
Oct	59	38	85	17		28	3	T	20	0	16	15	1	1	0	7	43	82	59	10	S	R-2
Nov	44	28	75	−3		14	3	6	16	2	9	21	0	1	0	18	26	81	68	12	S	R, S-2
Dec	31	15	67	−29		−2	2	13	17	4	10	21	0	1	0	28	25	80	69	11	S	S, R-2
Year	55	34	101	−30			32	64	214	19	167	198	27	11	4	154	46	78	60	10	S	

Average date of first freeze .... October 3
" " " last " .... May 8
" freeze-free period .... 148 days
10 inches of snow equal approximately one inch of rain.

Notes:

T Indicates "trace"

* For full explanation of (T-H-I) "Temperature Humidity Index;" "Wind Chill Factor" and "Storm Intensity," see beginning of Chapter 2.

55

CONCORD, NEW HAMPSHIRE — Elevation 339 Feet — Table 4

Month	Average Max.	Average Min.	Extreme Max.	Extreme Min.	T H I*	Wind Chill Factor*	Precipitation Total	Precipitation Snow	Not even 0.01" precip.	More than ½" of snow	Clear	Cloudy	Thunderstorms	Fog	90° or higher	32° or lower	% of possible sunshine	Rel. Humidity A.M.	Rel. Humidity P.M.	Wind M.P.H.	Wind Direction	Storm Intensity*
Jan	32	9	72	−35		−1	3	18	19	5	16	15	0	2	0	30	48	78	60	6	NW	S, R-2
Feb	33	10	68	−37		−3	2	17	18	5	16	12	0	1	0	27	53	74	56	7	NW	S, R-2
March	43	21	85	−16		10	3	12	20	4	16	15	1	2	0	25	55	74	53	7	NW	S, R-2
April	56	30	92	7		21	3	4	18	1	15	15	1	2	0	14	52	72	47	7	NW	R, S-2
May	69	41	98	21			3	T	18	0	16	15	3	3	1	1	51	75	48	6	NW	R-2
June	78	50	101	32	72		4	0	19	0	18	12	5	5	2	0	56	76	49	5	NW	T-3
July	83	55	102	38	76		4	0	21	0	20	11	6	7	3	0	56	80	49	5	NW	T-3
Aug	80	53	100	33	74		3	0	22	0	20	11	5	10	2	0	58	85	53	5	NW	T-3
Sept	72	45	98	20			3	0	21	0	18	12	2	9	1	1	55	88	53	5	NW	R-2
Oct	62	34	92	16		30	3	T	22	0	19	12	1	7	0	8	51	86	52	5	NW	R-2
Nov	48	26	80	−17		18	4	6	19	1	15	15	0	3	0	20	43	84	59	6	NW	R, S-2
Dec	34	14	65	−24		5	3	12	21	4	17	14	0	2	0	28	43	79	59	6	NW	S, R-2
Year	57	32	102	−37			37	68	238	20	206	159	24	53	9	154	52	79	53	6	NW	

Notes:

T Indicates "trace"

* For full explanation of (T-H-I) "Temperature Humidity Index;" "Wind Chill Factor" and "Storm Intensity," see beginning of Chapter 2.

Average date of first freeze .... September 3
" " last " .... May 11
" freeze-free period .... 142 days
10 inches of snow equal approximately one inch of rain.

BOSTON, MASSACHUSETTS Elevation 15 Feet Table 5

Month	Temperatures Average Max.	Average Min.	Extreme Max.	Extreme Min.	THI*	Wind Chill Factor*	Precipitation Total	Snow	Not even 0.01" precip.	More than ½" of snow	Clear	Cloudy	Thunder-storms	Fog	90° or higher	32° or lower	% of possible sunshine	Rel. Hum. A.M.	Rel. Hum. P.M.	Wind M.P.H.	Direction	Storm Intensity*
Jan	37	23	62	−4		−1	4	13	18	3	15	16	0	2	0	25	53	68	58	15	NW	R, S-3
Feb	37	23	58	−3		−2	3	11	16	3	15	13	0	2	0	25	57	67	57	15	WNW	R, S-3
March	45	31	66	6		10	4	8	19	2	16	15	0	2	0	18	57	68	59	15	NW	R, S-3
April	56	40	82	22		24	4	1	17	0	15	15	1	2	0	2	56	68	55	14	WNW	R-2
May	68	50	93	37			3	T	18	0	17	14	2	3	0	0	58	69	55	13	SW	R-2
June	76	59	94	46	72		3	0	20	0	17	13	4	2	3	0	63	74	60	12	SW	T-3
July	82	65	98	54	74		3	0	21	0	19	12	5	2	3	0	65	73	54	11	SW	T-3
Aug	80	63	93	47	74	22	4	0	21	0	20	11	4	2	2	0	65	75	58	11	SW	T-3
Sept	73	57	92	38			3	0	21	0	19	11	2	2	0	0	64	79	60	12	SW	R-2
Oct	63	47	85	32			3	T	21	0	19	12	1	2	0	1	61	75	55	13	SW	R-2
Nov	52	38	69	21		22	4	1	19	0	16	14	0	2	0	8	52	74	62	13	SW	R-2
Dec	40	27	70	−3		4	4	7	21	2	17	14	0	1	0	20	54	70	59	14	WNW	R, S-3
Year	59	44	98	−4			43	42	232	10	205	160	19	23	8	99	60	72	58	13	SW	

Notes:

T Indicates "trace"

* For full explanation of (T-H-I) "Temperature Humidity Index," "Wind Chill Factor" and "Storm Intensity," see beginning of Chapter 2.

Average date of first freeze .... October 25
" " " last " ....April 16
" freeze-free period ......191 days
10 inches of snow equal approximately one inch of rain.

57

NANTUCKET, MASSACHUSETTS Elevation 43 Feet Table 6

Month	Temperatures Average Max.	Average Min.	Extreme Max.	Extreme Min.	T H I *	Wind Chill Factor*	Precipitation Total	Precipitation Snow	Not even 0.01" precip.	More than ½" of snow	Clear	Cloudy	Thunder-storms	Fog	90° or higher	32° or lower	% of possible sunshine	Rel. Hum. A.M.	Rel. Hum. P.M.	Wind M.P.H.	Wind Direction	Storm Intensity*
Jan	39	27	63	2		1	4	9	17	2	12	19	0	5	0	25	42	80	67	15	NW	R, S-3
Feb	38	25	54	5		0	4	11	17	2	12	16	0	5	0	26	49	78	65	15	WNW	R, S-3
March	43	30	60	10		8	5	7	18	2	16	15	1	6	0	20	56	82	69	15	NW	R, S-3
April	51	38	69	23		20	4	1	18	0	14	16	2	8	0	11	56	80	66	15	WSW	R-2
May	60	45	77	32			3	0	20	0	14	17	3	10	0	1	59	82	71	13	SW	R-2
June	68	55	86	40			3	0	22	0	15	15	3	12	0	0	62	90	78	12	SW	R-2
July	74	62	90	50	71		3	0	23	0	16	15	4	15	0	0	61	88	76	11	SW	R-2
Aug	74	62	86	46	72	22	4	0	22	0	17	14	3	13	0	0	60	91	77	11	SW	R-2
Sept	69	56	83	36			4	0	23	0	17	13	2	7	0	0	60	89	72	12	SW	R-2
Oct	61	48	77	22			4	T	23	0	17	14	1	7	0	1	58	85	68	13	SW	R-3
Nov	52	40	68	20		22	4	T	18	0	14	16	1	5	0	8	42	81	72	13	NW	R-3
Dec	43	30	58	4		7	4	7	19	2	15	16	0	4	0	20	42	79	69	14	WNW	R, S-3
Year	56	43	90	2			44	35	240	8	179	186	20	98	0	112	55	84	71	13	SW	

Notes:
T Indicates "trace"
* For full explanation of (T-H-I) "Temperature Humidity Index," "Wind Chill Factor" and "Storm Intensity," see beginning of Chapter 2.

Average date of first freeze . . . . November 16
　　" 　" last 　" 　 . . . . April 12
　　" 　freeze-free period 　 . . . . 219 days
10 inches of snow equal approximately one inch of rain.

HARTFORD, CONNECTICUT Elevation 169 Feet Table 7

Month	Average Max.	Average Min.	Extreme Max.	Extreme Min.	T·H·I*	Wind Chill Factor*	Precip. Total	Precip. Snow	Not even 0.01" precip.	More than ½" of snow	Clear	Cloudy	Thunder-storms	Fog	90° or higher	32° or lower	% of possible sunshine	R.H. A.M.	R.H. P.M.	Wind M.P.H.	Wind Direction	Storm Intensity*
Jan	36	18	70	−17		0	3	11	20	3	16	15	0	3	0	29	46	73	61	9	N	R, S-3
Feb	38	18	72	−24		0	3	12	18	3	13	15	0	2	0	25	55	73	59	9	N	R, S-3
March	47	27	86	−4		9	4	7	19	4	16	15	1	2	0	20	56	72	55	10	N	R, S-3
April	60	36	91	11	22		4	1	16	1	14	16	2	2	0	6	54	72	52	10	S	R-2
May	72	47	94	28			4	T	19	0	17	14	3	3	0	0	57	71	52	9	S	R-2
June	81	57	100	38	75		4	0	19	0	16	14	5	2	2	0	60	73	56	8	S	T-3
July	86	62	101	48	78		4	0	20	0	18	13	7	2	4	0	62	76	55	8	S	T-3
Aug	83	60	101	38	76		4	0	20	0	19	12	5	5	3	0	60	78	55	7	S	T-3
Sept	76	52	101	30	72		3	0	20	0	17	13	2	8	1	0	57	82	56	7	S	R-2
Oct	65	41	91	18			3	T	22	0	15	13	1	7	0	3	55	84	54	8	N	R-2
Nov	51	31	83	6		17	3	2	19	0	16	14	0	4	0	14	46	79	58	9	S	R-2
Dec	39	20	67	−18		14	3	8	21	3	16	15	0	3	0	25	45	76	60	9	NW	R, S-3
Year	61	39	101	−24			40	40	233	14	196	169	28	43	10	120	55	77	56	8	S	

Average date of first freeze .... October 19
    "   "   last .... April 22
    "   freeze-free period .... 180 days
10 inches of snow equal approximately one inch of rain.

Notes:
T Indicates "trace"
* For full explanation of (T-H-I) "Temperature Humidity Index;" "Wind Chill Factor" and "Storm Intensity," see beginning of Chapter 2.

PROVIDENCE, RHODE ISLAND Elevation 55 Feet Table 8

Month	Average Max.	Average Min.	Extreme Max.	Extreme Min.	THI*	Wind Chill Factor*	Precip. Total	Precip. Snow	Not even 0.01" precip.	More than ½" of snow	Clear	Cloudy	Thunder-storms	Fog	90° or higher	32° or lower	% of possible sunshine	Rel. Hum. A.M.	Rel. Hum. P.M.	Wind M.P.H.	Wind Direction	Storm Intensity*
Jan	37	21	68	−9		2	4	10	19	3	16	15	0	3	0	26	49	73	60	12	NW	R, S-3
Feb	37	20	69	−17		0	3	10	18	2	15	13	0	2	0	24	56	73	56	12	NW	R, S-3
March	45	29	90	2		10	4	6	19	3	18	13	1	2	0	18	58	72	53	13	NW	R, S-3
April	55	37	91	11		22	3	1	17	0	16	14	1	3	0	5	57	71	51	12	SW	R-2
May	66	47	95	29			3	T	19	0	18	13	3	3	0	0	60	74	55	11	SSW	R-2
June	75	56	101	39	70		3	0	19	0	18	12	4	3	2	0	63	76	56	10	SW	R-2
July	80	62	101	49	75		3	0	21	0	19	12	5	2	4	0	63	79	56	9	SW	R-2
Aug	79	60	102	44	74		4	0	21	0	19	12	4	2	2	0	62	82	56	9	SW	R-2
Sept	72	53	99	33			3	0	21	0	18	12	2	3	1	0	60	84	56	9	SW	R-2
Oct	62	43	90	25		20	3	T	22	0	19	12	1	4	0	1	60	83	54	10	SW	R-2
Nov	51	34	82	9		20	4	1	18	0	16	14	0	3	0	11	51	80	58	11	SW	R-2
Dec	39	24	68	−12		8	3	6	20	2	17	14	0	2	0	23	50	74	57	11	WNW	R, S-3
Year	58	41	102	−17			40	33	234	10	209	156	21	32	9	108	58	77	56	11	SW	

Average date of first freeze .... October 27
     "   "   last   " .... April 13
     "   freeze-free period .... 197 days
10 inches of snow equal approximately one inch of rain.

Notes:
T Indicates "trace"
* For full explanation of (T-H-I) "Temperature Humidity Index," "Wind Chill Factor" and "Storm Intensity," see beginning of Chapter 2.

ALBANY, NEW YORK Elevation 227 Feet Table 9

Month	Avg Max	Avg Min	Extreme Max	Extreme Min	THI*	Wind Chill Factor*	Precip Total	Snow	Not even 0.01" precip.	More than ½" of snow	Clear	Cloudy	Thunderstorms	Fog	90° or higher	32° or lower	% of possible sunshine	Rel. Hum. A.M.	Rel. Hum. P.M.	Wind M.P.H.	Wind Direction	Storm Intensity*
Jan	31	14	64	−26		−3	2	16	18	4	13	18	0	1	0	29	42	75	63	10	WNW	S, R-2
Feb	32	14	63	−22		−6	2	13	17	4	13	15	0	1	0	27	51	75	61	11	WNW	S, R-2
March	42	24	85	−21		6	2	10	19	3	15	16	1	2	0	25	53	76	56	11	WNW	S, R-2
April	56	35	93	14		22	3	3	17	1	13	17	2	1	0	10	52	73	50	10	WNW	R-2
May	69	46	92	27			3	0	18	0	14	17	4	2	0	1	57	73	51	9	S	R-2
June	79	55	99	33	73		3	0	20	0	17	13	7	1	4	0	62	75	52	8	S	T-3
July	83	60	100	44	76		3	0	20	0	20	11	8	1	6	0	65	79	52	7	S	T-3
Aug	81	58	99	35	75		3	0	21	0	20	11	5	3	4	0	61	83	52	6	S	T-3
Sept	73	50	100	24			3	0	21	0	19	11	3	3	1	1	58	87	55	7	S	R-2
Oct	62	40	91	19		31	2	0	22	0	18	13	1	4	0	7	56	86	54	8	S	R-2
Nov	47	31	82	−11		19	3	4	19	1	13	17	0	2	0	18	38	81	61	9	S	R-2
Dec	34	19	62	−19		5	2	10	20	3	13	18	0	2	0	27	37	78	64	9	S	S, R-2
Year	57	37	100	−26			32	56	232	16	188	177	31	23	15	145	54	78	56	9	S	

Average date of first freeze . . . . October 16
" " last " . . . . April 24
" freeze-free period . . . . 175 days
10 inches of snow equal approximately one inch of rain.

Notes:

T Indicates "trace"

* For full explanation of (T-H-I) "Temperature Humidity Index," "Wind Chill Factor" and "Storm Intensity;" see beginning of Chapter 2.

NEW YORK, NEW YORK — Elevation 10 Feet — Table 10

Month	Temperatures Average Max.	Average Min.	Extreme Max.	Extreme Min.	THI*	Wind Chill Factor*	Precip. Total	Precip. Snow	Not even 0.01" precip.	More than ½" of snow	Clear	Cloudy	Thunder-storms	Fog	90° or higher	32° or lower	% of possible sunshine	Rel. Hum. A.M.	Rel. Hum. P.M.	Wind M.P.H.	Wind Direction	Storm Intensity*
Jan	40	26	71	-6		0	4	7	19	2	18	13	0	3	0	24	51	72	62	16	NW	R, S-3
Feb	40	25	73	-14		-3	3	9	18	2	17	11	0	2	0	22	57	70	59	17	NW	R, S-3
March	49	33	84	3		13	4	6	19	2	19	12	1	2	0	16	58	69	56	17	NW	R, S-3
April	58	42	91	12			3	1	19	0	19	11	2	2	0	4	60	68	54	15	NW	R-2
May	69	53	95	34			4	T	20	0	20	11	4	2	1	0	61	71	56	14	NW	R-2
June	78	62	97	44	73		4	0	20	0	21	9	6	1	3	0	64	74	58	13	S	T-3
July	82	67	102	54	76		4	0	20	0	21	10	7	1	2	0	65	75	57	12	S	T-3
Aug	80	66	102	51	75		4	0	21	0	21	10	6	0	1	0	63	77	60	12	S	T-3
Sept	75	60	100	39	71		4	0	21	0	21	9	3	1	0	0	64	78	60	12	N	R-2
Oct	65	50	92	27			3	T	22	0	21	10	1	2	0	0	63	75	58	14	NW	R-2
Nov	53	40	81	7		21	3	1	21	0	18	11	1	2	0	6	56	73	61	16	NW	R-2
Dec	42	29	69	13		25	3	6	21	2	18	13	0	2	0	20	53	72	62	16	NW	R, S-3
Year	61	46	102	-14			42	30	241	8	135	130	31	20	7	92	60	73	59	15	NW	

Average date of first freeze . . . . November 24
" " last " . . . . April 2
" freeze-free period . . . . 236 days
10 inches of snow equal approximately one inch of rain.

Notes:
T Indicates "trace"
* For full explanation of (T-H-I) "Temperature Humidity Index;" "Wind Chill Factor" and "Storm Intensity;" see beginning of Chapter 2.

NEWARK, NEW JERSEY

Elevation 11 Feet

Table 11

Month	Avg Max	Avg Min	Extreme Max	Extreme Min	THI*	Wind Chill Factor*	Precip Total	Precip Snow	Not even 0.01" precip.	More than ½" of snow	Clear	Cloudy	Thunderstorms	Fog	90° or higher	32° or lower	% of possible sunshine	RH A.M.	RH P.M.	Wind M.P.H.	Wind Direction	Storm Intensity*
Jan	40	25	74	0		8	3	8	19	2	14	17	0	3	0	25	50	72	58	11	NE	R, S-3
Feb	41	25	76	−7		6	3	8	18	2	15	13	0	2	0	22	50	70	56	12	NW	R, S-3
March	49	32	89	6		17	4	5	19	1	17	14	1	2	3	14	53	66	52	12	NW	R, S-3
April	61	42	91	23		27	4	T	18	0	16	14	1	1	0	2	54	69	49	11	WNW	R-2
May	72	52	98	33	75		4	T	18	0	17	14	4	2	1	0	60	71	51	10	SW	R-2
June	81	61	102	43	78		3	0	20	0	19	11	5	2	6	0	62	73	52	9	SW	T-3
July	86	67	105	52	77		4	0	21	0	19	12	6	1	9	0	62	74	51	9	SW	T-3
Aug	84	65	103	51	72		4	0	22	0	19	12	4	1	6	0	61	78	53	9	SW	T-3
Sept	77	58	105	35			4	0	21	0	18	12	2	1	2	0	60	80	53	9	SW	R-3
Oct	66	47	92	30			3	T	23	0	19	12	1	3	0	0	60	79	52	9	SW	R-2
Nov	54	37	85	15		25	3	T	20	0	17	13	0	2	0	8	50	76	55	10	SW	R-2
Dec	42	27	72	−1		11	3	8	20	2	17	14	0	2	0	22	49	73	58	11	SW	R, S-3
Year	63	45	105	−7			42	29	239	7	207	158	25	22	25	94	59	73	53	10	SW	

Notes:

T Indicates "trace"

* For full explanation of (T-H-I) "Temperature Humidity Index;" "Wind Chill Factor" and "Storm Intensity," see beginning of Chapter 2.

Average date of first freeze . . . . November 2

"   "   "   last   "   . . . . April 7

"   "   freeze-free period . . . . 208 days

10 inches of snow equal approximate one inch of rain.

ATLANTIC CITY, NEW JERSEY — Elevation 10 Feet — Table 12

Month	Average Max.	Average Min.	Extreme Max.	Extreme Min.	T H I *	Wind Chill Factor*	Precip. Total	Precip. Snow	Not even 0.01" precip.	More than ½" of snow	Clear	Cloudy	Thunder-storms	Fog	90° or higher	32° or lower	% of possible sunshine	Rel. Hum. A.M.	Rel. Hum. P.M.	Wind M.P.H.	Wind Direction	Storm Intensity*
Jan	43	27	65	−8		6	4	6	19	1	17	14	0	3	0	28	58	74	58	13	WNW	R, S-3
Feb	43	26	69	0		6	3	4	18	2	18	10	0	5	0	21	52	80	57	12	W	R, S-3
March	50	32	67	13		16	4	3	19	1	19	12	1	4	0	21	56	78	51	13	WNW	R, S-3
April	60	42	85	24			3	T	19	0	19	11	2	3	0	8	50	74	52	12	S	R-2
May	71	52	93	25			4	0	20	0	21	10	3	3	2	1	59	77	53	11	S	R-2
June	79	61	96	39	72		3	0	20	0	21	9	5	4	7	0	62	79	51	10	S	T-3
July	84	66	104	46	76		4	0	21	0	22	9	6	4	7	0	65	82	51	10	S	T-3
Aug	82	65	95	40	76		5	0	21	0	22	9	4	4	4	0	62	86	55	9	S	T-3
Sept	76	58	92	35	71		3	0	22	0	21	9	2	3	2	0	60	87	59	10	ENE	R-2
Oct	67	48	78	26			3	T	22	0	21	10	1	5	0	4	65	86	51	11	W	R-2
Nov	56	38	76	11			4	T	21	0	19	11	0	3	0	15	58	83	55	12	WNW	R-2
Dec	45	28	72	6		11	3	3	21	1	20	11	0	5	0	23	48	83	57	11	WNW	R, S-3
Year	63	45	104	−8			42	16	243	5	240	125	24	44	22	120	58	80	54	11	S	

Notes:

T Indicates "trace"

* For full explanation of (T-H-I) "Temperature Humidity Index," "Wind Chill Factor" and "Storm Intensity," see beginning of Chapter 2.

Average date of first freeze .... November 7

" " last " .... April 4

" freeze-free period .... 217 days

10 inches of snow equal approximately one inch of rain.

64

PHILADELPHIA, PENNSYLVANIA   Elevation 7 Feet   Table 13

Month	Average Max.	Average Min.	Extreme Max.	Extreme Min.	THI*	Wind Chill Factor*	Precip. Total	Precip. Snow	Not even 0.01" precip.	More than ½" of snow	Clear	Cloudy	Thunder-storms	Fog	90° or higher	32° or lower	% of possible sunshine	R.H. A.M.	R.H. P.M.	Wind M.P.H.	Wind Direction	Storm Intensity*
Jan	41	25	74	2		11	3	5	19	2	14	17	0	4	0	24	46	77	60	10	WNW	R, S-3
Feb	42	25	74	3		9	3	5	19	1	15	13	0	3	0	21	55	75	56	11	NW	R, S-3
March	52	33	87	7		18	3	3	19	1	16	15	1	2	0	14	57	74	52	12	WNW	R, S-3
April	62	41	92	24			3	T	19	0	15	15	2	2	0	2	56	73	48	11	SW	R-2
May	74	52	96	33			4	0	19	0	16	15	5	2	1	0	58	75	51	10	WSW	R-2
June	83	62	100	44	77		4	0	20	0	19	11	6	2	7	0	63	77	52	9	SW	T-3
July	87	66	102	52	79		4	0	21	0	19	12	6	1	11	0	64	78	51	8	WSW	T-3
Aug	84	64	101	50	77		5	0	21	0	19	12	5	2	6	0	61	81	54	8	SW	T-3
Sept	78	58	100	36	73		3	0	22	0	19	11	2	2	2	0	58	83	52	8	SW	R-2
Oct	67	46	96	28		25	3	T	23	0	19	12	1	5	0	1	58	83	52	9	SW	R-2
Nov	55	37	79	15			3	1	21	0	17	13	1	3	0	9	53	80	55	10	WNW	R-2
Dec	44	27	72	1		13	3	3	22	1	17	14	0	4	0	21	50	77	58	10	WNW	R-3
Year	64	45	102	1			41	17	245	5	205	160	29	32	27	92	57	78	53	10	SW	

Notes:

T Indicates "trace"

* For full explanation of (T-H-I) "Temperature Humidity Index," "Wind Chill Factor" and "Storm Intensity," see beginning of Chapter 2.

Average date of first freeze .... November 17

" " last " .... March 30

" freeze-free period .... 232 days

10 inches of snow equal approximately one inch of rain.

65

In most of inland Massachusetts, afternoon highs average in the upper seventies. Albany, New York, although in the same latitude, registers in the mid eighties during the day. In these sections, the evenings cool considerably with the mercury dropping to the upper fifties and lower sixties. The months of June and the latter part of August are somewhat cooler.

Beaches in Connecticut, Rhode Island, Long Island and New Jersey almost always ensure temperatures in the low to mid seventies and pleasant afternoon sea breezes. Especially inviting are lovely Cape Cod and Nantucket Island where temperatures in the afternoon are in the mid-seventies and T.H.I.'s in the comfortable low seventies. Evenings are in the upper fifties and low sixties. Heat waves which bring readings of 100 degrees inland cannot push above the low to mid nineties on the Cape. Throughout the summer there is much sunshine. Table 1 lists the percent of possible sunshine at a number of locations. On the detailed map, chart no. 17, included with this area write-up is shown the total number of hours of sunshine per year at most of these places.

THE MID-ATLANTIC STATES—Area Number 2

See Chart No. 19

(Kentucky, Tennessee, North Carolina, South Carolina, Virginia, Maryland, Delaware)

This entire area ranks very high in popularity with vacationists, and many portions are also rapidly becoming major retirement centers. Enjoying the stimulation of definite change of seasons, it still escapes most of the violent and stormy weather experienced in the north. Conversely, although these states lie below the Mason-Dixon Line, they generally suffer somewhat less from the extremely oppressive humidity than many places farther south.

There is a wide choice of climate and living conditions, from the low tidal flats to the Blue Ridge and Smoky Mountain high country. And there are many choice spots in between. One such is North Carolina's fast growing "research triangle," the home of artists, writers and all manner of retirees, who relish the quiet, cultural atmosphere. If the heat of summer becomes a bit irksome, a few hours will get them to either

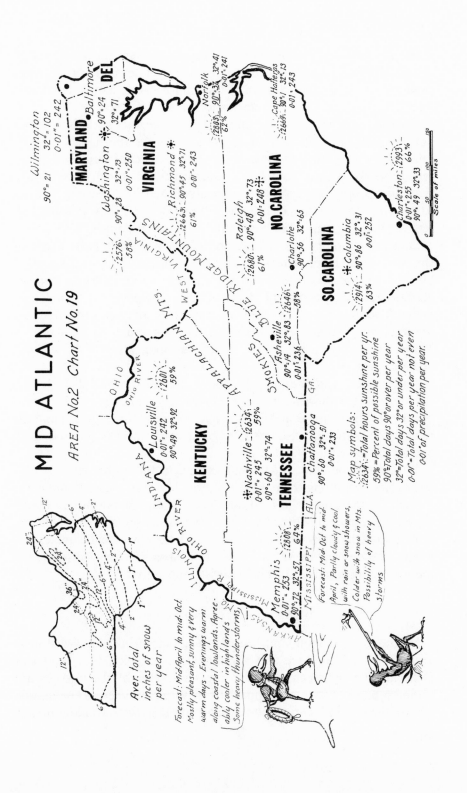

MID ATLANTIC

AREA No.2 Chart No.19

DEL

MARYLAND Baltimore ✳ 90°·24 32°·71 0·01"·242

Wilmington 90°·21 32°·102 0·01"·242

Washington ✳ 90°·28 32°·73 0·01"·250

VIRGINIA ✳

Richmond ✳ :2663: 90°·45 32°71 61% 0·01"·243

Norfolk :2803: 90°·3 32°·41 62% 0·01"·241

Cape Hatteras :2669: 90°·1 32°·13 0·01"·243

NO.CAROLINA

Raleigh :2680: 90°·48 32°·73 61%

Charlotte 90°·56 32°·65

:2576: 58%

WEST VIRGINIA

BLUE RIDGE MOUNTAINS

SO.CAROLINA

Columbia ✳ :2914: 90°·86 32°·31 63% 0·01"·252

Charleston :2993: 66% 0·01"·255 90°·49 32°·33

Scale of miles

APPALACHIAN MTS.

SMOKIES BLUE

Asheville :2646: 90°·14 32°·83 58% 0·01"·236

GA.

OHIO

OHIO RIVER

:260: 59%

Louisville :2634: 0·01"·242 90°·49 32°·92 59%

INDIANA

ILLINOIS

OHIO RIVER

KENTUCKY

TENNESSEE

✳ Nashville :2634: 0·01"·245 90°·60 32°·74

Chattanooga 90°·80 32°·5 0·01"·233

ALA.

Memphis :2808: 64% 0·01"·253 90°·72 32°·57

MISSISSIPPI

ARKANSAS

MISSISSIPPI R.

Aver. total
inches of snow
per year

Forecast: Mid-April to mid-Oct.
Mostly pleasant, sunny & very
warm days - Evenings warm
along coastal lowlands. Agreeably cooler in highlands.
Some heavy thunderstorms

Forecast: Mid-Oct. to mid-
April. Partly cloudy & cool
with rain or snow showers.
Colder with snow in Mts.
Possibility of heavy
storms

Map symbols:
:2634: =Total hours sunshine per yr.
59%=Percent of possible sunshine
90°=Total days 90°or over per year
32°=Total days 32°or under per year
0·01"=Total days per year not even
0·01"of precipitation per year.

the cooler mountains or the seashore breezes. Perhaps the most popular region with both vacationists and retirees is the tidewater of the Chesapeake Bay. The overall climate of this mid-Atlantic area is temperate and relatively free from extremes and sudden changes.

Spring and autumn are generally the prime weather periods throughout all of these states. Although there is always the possibility of a hurricane in late summer and early fall, the chances of actually encountering one are quite remote. So few hurricanes ever touch these shores, that most people who have lived here all of their lives, never get closer to one than the newspaper headlines. The efficient ESSA warning system greatly minimizes the possibility of personal danger.

Summer humidity can be high and often oppressive, particularly in the low coastal lands. The thermometer can also reach or even top 100°, but that is most noticeable in the middle country. At the higher elevations and immediately along the shore, breezes are a welcome factor. The summer heat-humidity combination of a few spots (like the lower parts of South Carolina) may be a bit much for a native of Minnesota, but those who have retired there soon adjust to the weather and way of life.

The frequent and occasionally violent thundershowers also offer refreshing relief as they can drop the temperature 5 or even 10°. Throughout the spring, summer and fall seasons visitors come in great numbers to enjoy the sunshine, fishing, golfing and every form of outdoor, land and water recreational activities.

Winter can be cold, snowy and snappy in the high country, cold, dreary and wet along the northern shores, but very pleasant and comfortable in the great mid-land stretches. Also that hot-humid South Carolina summer area becomes a welcome winter refuge for great flocks of migratory humans. If there is no such thing as an ideal climate, this region of wide weather variety seems to be a satisfactory substitute for the increasing numbers of both tourists and retirees.

Winter Temperatures

Maximum afternoon temperatures in January usually hover around a mild 50° in Tennessee and Virginia. In Kentucky, Maryland and Delaware it may reach only the low 40s. At night the low 30s are common in Tennessee and upper 20s in Virginia, whereas the middle 20s are the rule in Kentucky, Maryland and Delaware. Cold outbreaks from Canada can drop temperatures well below zero in these areas and below 10° in Tennessee. February and December are a few degrees warmer than January. Daytime January readings are generally in the low 50s in North Carolina and comfortable upper 50s in South Carolina. Places in the higher western parts of the states register several degrees lower. At night temperatures drop into the low 30s in North Carolina and mid to upper 30s in South Carolina. December and February are slightly warmer than January but cold waves may bring the low teens to the inland areas.

Moderate precipitation falls in this area, much of it being rain. Storms moving along the Gulf states and then up the Atlantic coast spread moderate rains and sometimes snow over the entire area. During the winter months, an average of 4″ of snow per month can fall in most of Kentucky. Tennessee and Virginia, which experience more rain, average only about 2″ of snow per month. Occasionally, however, heavy snowstorms do strike, dumping 10″ of snow or more. That, however, is rare. In eastern areas, around the Cumberland Plateau and the Appalachian Mountains, the precipitation is somewhat heavier. This is especially true toward the extreme Southeastern corner, in the vicinity of Great Smoky National Park. In Maryland and Delaware, 4″ to 6″ of snow per month can be expected. An occasional heavy snowstorm is possible here, especially during late January and February. In North and South Carolina most precipitation is in the form of rain. Inland mountain areas occasionally experience snow—especially during January and February, which may average 2 to 3 inches per month. As much as 15″ of snow, however, can fall during one severe snowstorm, especially in western North Carolina. This rarely occurs more than once a year. Coastal areas receive most of their rainfall from developing coastal storms which originate in the Cape Hatteras area. Moderate to heavy rains can accompany these storms.

SPRING AND FALL

Fall temperature readings are slightly higher than corresponding spring ones. In March afternoons may reach the low 50s in Kentucky, Delaware and Maryland and near 60° in Tennessee and Virginia, with slightly cooler weather in the hilly areas. At night temperatures drop to the freezing mark in Kentucky, Delaware and Maryland, and into the mid to upper 30s in Tennessee and Virginia. They go up about 10 degrees per month during April and May, making May very pleasant. Maximum temperatures are in the mid 70s throughout Delaware, Maryland and coastal Virginia, with low 80s in Tennessee. Nights are comfortable, being in the mid to upper 50s. September and October are ideal times to visit the mountains because temperatures reach the low to mid 80s throughout the entire area. Comfortable sleeping weather prevails as the mercury drops to the pleasant 50s and 60s at night. November brings a chill to the air and the chance of some light snow, especially at the higher elevations in northern areas.

The Carolina coastal regions have a smaller temperature range than do inland areas; consequently nights are not as chilly. During March, inland North Carolina experiences readings in the mid to upper 30s at night, while those along the coast remain in the mid 40s. The same is true in South Carolina with inland night lows of about 40 while those at the coast are nearer 50. Daytime temperatures generally climb to the upper 50s and low 60s during March in North Carolina, and mid 60s in South Carolina, April is about 8 to 10 degrees warmer than March. May registers figures in the upper 70s in North Carolina and low to mid 80s in South Carolina. The ocean has little effect on the maximum temperatures.

Spring is rather wet but the fall is dry. Snow threatens the area, especially the northern and inland sections until early March. About 4 inches will usually fall in Maryland and Delaware, and up to 9 inches in the mountains of Virginia. There will be only 1 to 2 inches of snow along coastal Virginia, with mostly rain elsewhere. April can be a month of rain and windy weather. Thunderstorms also become more numerous and there may be as many as two a week in the mountainous country.

Some of these storms can become quite strong, but the possibility of tornadoes is small. Rainfall is heavy but usually of short duration. Fall is characterized by fine weather with mostly sunny days. Rain, when it occurs, is usually light and brief but of longer duration than spring. Thunderstorms become less severe and less numerous. Snow can appear by early November, especially in the higher elevations.

Table 14 lists the percentage of possible sunshine at a number of locations in this area and the detailed map of these states (see chart no. 19) shows the total number of hours of sunshine per year at most of these places.

Spring precipitation in the Carolinas is usually in the form of rain; however snow, especially in inland areas, is very probable in March during severe eastcoast storms. Coastal areas generally have rain, and it is rare to see snow during April and May, even in the inland mountains. Thunderstorms are on the increase during these months, especially in May, and one to two thunderstorms per week are not unusual, particularly in South Carolina. The major weather hazard during the fall months is the threat of hurricanes. Almost every year these tropical storms threaten the Carolinas, but only about one actually strikes every two or three years and then usually just the eastern tip of North Carolina which juts out into the Atlantic at Hatteras. Moderate rainstorms also threaten the Carolinas during the fall. Four to six inches of rain is usual along the coast during September, with lesser amounts inland. The frequency of heavy rainstorms diminishes as the fall season progresses.

Summer

Summers are moderately hot, especially in western Tennessee. Afternoon temperatures can be expected to soar into the low 90s on most days and this, coupled with moderately high humidities, causes many

Table 14	AVERAGE % OF MAXIMUM POSSIBLE SUNSHINE (% of daylight hours)												
Area Number 2 Mid-Atlantic	Jan	Feb	March	April	May	June	July	Aug	Sept	Oct	Nov	Dec	Year
Virginia													
Norfolk	50	57	60	63	67	66	66	66	63	64	60	51	62
Richmond	49	55	59	63	67	66	65	62	63	64	58	50	61
North Carolina													
Asheville	48	53	56	61	64	63	59	59	62	64	59	48	58
Raleigh	50	56	59	64	67	65	62	62	63	64	62	52	61
South Carolina													
Charleston	58	60	65	72	73	70	66	66	67	68	68	57	66
Columbia	53	57	62	68	69	68	63	65	64	68	64	51	63
Kentucky													
Louisville	41	47	52	57	64	68	72	69	68	64	51	39	59
Tennessee													
Knoxville	42	49	53	59	64	66	64	59	64	64	53	41	57
Memphis	44	51	57	64	68	74	73	74	70	69	58	45	64
Nashville	42	47	54	60	65	69	69	68	69	65	55	42	59

WASHINGTON D.C.

Elevation 14 Feet **Table 15**

Month	Avg Max	Avg Min	Extreme Max	Extreme Min	THI*	Wind Chill Factor*	Precip Total (in)	Snow (in)	Days not even 0.01" precip	Days more than ½" snow	Clear	Cloudy	Thunderstorms	Fog	90° or higher	32° or lower	% of possible sunshine	RH A.M.	RH P.M.	Wind M.P.H.	Wind Direction	Storm Intensity*
Jan	44	30	71	3		13	3	5	20	2	14	17	0	2	0	24	50	69	54	10	NW	R, S-3
Feb	46	29	71	4		13	2	6	20	1	14	14	0	2	0	20	51	70	56	11	S	R, S-3
March	54	36	86	16		22	3	3	19	1	17	14	1	1	0	9	56	70	49	11	NW	R, S-3
April	66	46	89	30			3	T	20	0	15	15	3	1	0	2	54	69	48	11	S	R-3
May	76	56	96	36	69		4	T	19	0	17	14	5	1	3	0	57	72	49	9	S	T-3
June	83	65	100	47	77		3	0	21	0	20	10	5	0	9	0	66	75	51	9	S	T-3
July	87	69	101	56	78		4	0	21	0	20	11	6	0	13	0	63	76	51	8	S	T-3
Aug	85	68	99	51	77		5	0	21	0	19	12	5	0	9	0	61	79	53	8	S	T-3
Sept	79	61	95	39	73		4	0	22	0	19	11	2	1	4	0	63	79	55	9	S	R-2
Oct	68	50	91	30			3	T	23	0	19	12	1	2	0	1	62	79	49	9	SSW	R-2
Nov	57	39	81	21		27	3	1	22	0	17	13	0	2	0	5	53	72	50	9	S	R-2
Dec	46	31	72	10		17	3	4	22	1	16	15	0	3	0	20	49	72	57	9	NW	R, S-2
Year	66	48	101	3			41	19	243	5	207	158	28	14	38	81	57	73	52	10	S	

Notes:

T Indicates "trace"

* For full explanation of (T-H-I) "Temperature Humidity Index;" "Wind Chill Factor" and "Storm Intensity," see beginning of Chapter 2.

Average date of first freeze . . . . October 28
" " " last " . . . . April 10
" " freeze-free period . . . . 200 days
10 inches of snow equal approximately one inch of rain.

WILMINGTON, DELAWARE   Elevation 78 Feet   Table 16

Month	Average Max.	Average Min.	Extreme Max.	Extreme Min.	THI*	Wind Chill Factor*	Precip. Total	Precip. Snow	Not even 0.01" precip.	More than ½" of snow	Clear	Cloudy	Thunder-storms	Fog	90° or higher	32° or lower	% of possible sunshine	Rel. Hum. A.M.	Rel. Hum. P.M.	Wind M.P.H.	Wind Direction	Storm Intensity*
Jan	42	25	75	−4		13	4	6	19	2	14	17	0	5	0	25	47	77	62	9	NW	R, S-3
Feb	43	25	74	2		11	3	4	18	1	14	14	0	5	0	21	52	77	59	9	NW	R, S-3
March	53	32	86	10		18	4	3	18	1	17	14	2	4	0	16	54	75	54	11	NW	R, S-3
April	63	40	88	23		28	4	T	18	0	15	15	3	3	0	3	55	74	52	10	WNW	R-3
May	75	51	93	34	70		4	0	19	0	16	15	6	3	1	0	61	77	54	9	NW	T-3
June	83	60	99	45	77		4	0	21	0	19	11	6	3	6	0	64	78	53	8	S	T-3
July	87	65	102	50	79		5	0	22	0	19	12	5	2	11	0	64	79	51	8	NW	T-3
Aug	85	63	101	48	79		5	0	22	0	19	12	6	4	6	0	63	84	55	7	NW	T-3
Sept	79	57	100	37	74		4	0	22	0	18	12	3	4	1	0	61	86	55	8	NNW	T-3
Oct	67	45	91	26			3	T	22	0	18	13	1	5	0	1	60	85	54	8	NW	R-2
Nov	55	36	85	14		26	3	2	21	0	16	14	1	5	0	12	51	82	57	9	NW	R-2
Dec	44	26	71	6		14	3	3	21	1	16	15	0	4	0	22	50	78	60	9	NW	R, S-2
Year	65	44	102	−4			45	18	243	5	201	164	32	47	24	100	57	79	56	9	NW	

Average date of first freeze .... October 26
"   "   last .... April 18
"   freeze-free period .... 191 days
10 inches of snow equal approximately one inch of rain.

Notes:
T Indicates "trace"
* For full explanation of (T-H-I) "Temperature Humidity Index," "Wind Chill Factor" and "Storm Intensity," see beginning of Chapter 2.

BALTIMORE, MARYLAND — Elevation 148 Feet — Table 17

Month	Average Max.	Average Min.	Extreme Max.	Extreme Min.	T-H-I*	Wind Chill Factor*	Precip. Total	Precip. Snow	Not even 0.01" precip.	More than ½" of snow	Clear	Cloudy	Thunderstorms	Fog	90° or higher	32° or lower	% of possible sunshine	Rel. Hum. A.M.	Rel. Hum. P.M.	Wind M.P.H.	Wind Direction	Storm Intensity*
Jan	44	25	75	−7		10	3	6	20	3	16	15	0	4	0	25	52	73	57	10	WNW	R, S-3
Feb	46	26	76	−1		8	3	7	20	2	15	13	0	4	0	22	54	75	58	11	NW	R, S-3
March	54	33	85	6		17	4	6	19	3	17	14	1	3	0	16	56	73	51	12	WNW	R, S-3
April	66	43	94	20			4	T	19	0	16	14	2	2	0	3	54	73	50	12	WNW	R-3
May	76	53	98	32			4	T	20	0	18	13	4	2	2	0	58	76	51	10	W	T-3
June	84	61	100	42	77		3	0	20	0	21	9	6	1	7	0	65	78	52	9	WNW	T-3
July	87	66	102	52	79		4	0	23	0	22	9	6	1	11	0	67	81	53	9	W	T-3
Aug	85	65	102	48	79		5	0	20	0	20	11	5	1	8	0	62	84	56	9	W	T-3
Sept	79	58	99	35	74		3	0	22	0	20	10	2	2	3	0	62	86	55	9	S	R-2
Oct	68	46	92	25			3	T	24	0	20	11	1	3	0	2	62	83	52	10	NW	R-2
Nov	57	34	83	13		24	3	1	21	1	16	14	0	3	0	13	53	78	53	10	WNW	R-2
Dec	46	26	74	0		13	3	5	22	1	17	14	0	4	0	23	49	76	58	10	WNW	R, S-2
Year	66	45	102	−7			43	26	250	10	218	147	27	31	31	103	58	78	54	10	W	

Notes:

T Indicates "trace"

* For full explanation of (T-H-I) "Temperature Humidity Index," "Wind Chill Factor" and "Storm Intensity," see beginning of Chapter 2.

Average date of first freeze . . . . .November 17
     "   " last   . . . . .March 28
     " freeze-free period   .234 days
10 inches of snow equal approximately one inch of rain.

74

RICHMOND, VIRGINIA — Elevation 144 Feet — Table 18

Month	Average Max.	Average Min.	Extreme Max.	Extreme Min.	THI*	Wind Chill Factor*	Precip. Total	Precip. Snow	Not even 0.01" precip.	More than ½" of snow	Clear	Cloudy	Thunder-storms	Fog	90° or higher	32° or lower	% of possible sunshine	R.H. A.M.	R.H. P.M.	Wind M.P.H.	Wind Direction	Storm Intensity*
Jan	48	30	80	−1		18	3	4	21	1	14	17	0	3			49	86	60	8	S	R,S-3
Feb	49	30	82	−3		17	3	3	19	1	15	13	0	2			54	78	60	9	WSW	R,S-2
March	59	38	94	12		26	4	3	20	1	17	14	1	2			57	77	54	10	W	R,S-2
April	68	46	96	19			4	T	20	0	17	13	2	2			63	73	52	9	S	R-2
May	77	55	98	35	72		4	0	20	0	18	13	6	2			65	74	52	8	SSW	T-3
June	84	64	104	43	77		4	0	21	0	20	10	7	2			69	77	52	7	S	T-3
July	88	68	105	52	80		5	0	19	0	19	12	10	2			70	79	57	7	SSW	T-4
Aug	86	67	107	49	79		5	0	21	0	19	12	7	3			65	82	58	8	S	T-4
Sept	81	61	101	40	76		3	0	22	0	17	13	3	4			64	83	57	7	S	R-3
Oct	71	49	99	28			3	T	23	0	19	12	1	4			58	82	52	7	NNE	R-2
Nov	59	39	83	14		29	2	1	21	0	17	13	1	2			54	79	56	8	S	R-2
Dec	49	31	77	−2		21	3	2	22	1	17	14	0	3			52	81	63	8	SW	R,S-2
Year	68	48	107	−3			41	12	249	4	209	156	38	31			60	79	56	8	S	

Notes:

T Indicates "trace"

* For full explanation of (T-H-I) "Temperature Humidity Index," "Wind Chill Factor" and "Storm Intensity," see beginning of Chapter 2.

Average date of first freeze . . . . November 8
" " last " . . . . April 2
" freeze-free period . . . . 220 days
10 inches of snow equal approximately one inch of rain.

NORFOLK, VIRGINIA — Elevation 26 Feet — Table 19

Month	Average Max.	Average Min.	Extreme Max.	Extreme Min.	THI*	Wind Chill Factor*	Precip. Total	Precip. Snow	Not even 0.01" precip.	More than ½" of snow	Clear	Cloudy	Thunderstorms	Fog	90° or higher	32° or lower	% of possible sunshine	R.H. A.M.	R.H. P.M.	Wind M.P.H.	Wind Direction	Storm Intensity*
Jan	49	34	80	5		18	3	2	21	1	15	16	0	2			58	79	62	12	NE	R, S-2
Feb	50	34	82	2		18	3	1	18	0	14	14	1	3			60	77	56	12	NNE	R, S-2
March	58	40	92	14		25	4	1	19	0	16	15	2	2			60	77	56	12	N	R-2
April	66	48	95	23	70		3	T	19	0	18	12	3	2			64	74	53	12	SW	R-2
May	76	58	98	33	77		4	0	21	0	19	12	6	2			66	76	54	10	SW	T-3
June	83	66	102	49	77		4	0	22	0	20	10	7	2			70	79	58	9	SW	T-3
July	87	71	104	57	79		6	0	20	0	20	11	9	1			66	81	61	8	SW	T-4
Aug	85	70	105	56	79		5	0	19	0	20	11	8	2			62	82	62	9	SW	T-4
Sept	80	65	100	40	75		4	0	22	0	18	12	3	2			63	82	63	9	NE	R-3
Oct	70	55	94	31			3	0	23	0	18	13	2	3			58	81	59	11	NE	R-2
Nov	60	45	82	17			3	T	21	0	19	11	0	3			61	80	58	11	SW	R-2
Dec	51	36	76	5		20	3	2	23	1	17	14	0	2			62	79	60	11	SW	R, S-2
Year	68	52	105	2		20	45	6	248	2	214	151	41	26			62	79	58	10	SW	

Notes:

T Indicates "trace"

* For full explanation of (T-H-I) "Temperature Humidity Index," "Wind Chill Factor" and "Storm Intensity," see beginning of Chapter 2.

Average date of first freeze ....November 27
" " last " .... March 18
" freeze-free period .... 254 days
10 inches of snow equal approximately one inch of rain.

CHARLOTTE, NORTH CAROLINA — Elevation 736 feet — Table 20

Month	Avg Max	Avg Min	Ext Max	Ext Min	THI*	Wind Chill Factor*	Precip Total	Snow	Days not even 0.01" precip.	Days more than ½" snow	Clear	Cloudy	Thunder-storms	Fog	90° or higher	32° or lower	% of possible sunshine	Rel. Hum. A.M.	Rel. Hum. P.M.	Wind M.P.H.	Wind Direction	Storm Intensity*
Jan	51	34	74	4		21	4	2	21	1	16	15	1	3	0	21	57	79	55	8	SW	R, S-2
Feb	54	35	78	7		23	4	1	18	0	15	13	1	3	0	17	59	76	52	8	NE	R, S-2
March	60	40	86	18		30	4	1	19	0	17	14	2	2	0	9	64	79	48	9	SW	R, S-2
April	71	50	91	28			3	T	21	0	19	11	3	1	1	1	69	81	49	9	S	R-2
May	79	59	95	32	74		3	0	23	0	19	12	6	1	3	0	69	84	52	8	SW	T-3
June	88	67	99	46	80		4	0	21	0	19	11	8	1	7	0	71	87	57	7	SW	T-3
July	89	70	99	53	81		5	0	19	0	18	13	10	1	12	0	69	89	58	7	SW	T-4
Aug	88	70	100	53	80		4	0	21	0	21	10	8	2	10	0	71	90	58	7	S	T-3
Sept	82	64	94	39	77		3	0	23	0	17	13	3	2	3	0	68	90	55	7	NE	R-3
Oct	73	52	87	24			3	0	24	0	20	11	1	2	0	2	71	88	50	7	NNE	R-2
Nov	61	40	85	20		31	3	T	22	0	19	11	1	2	0	8	63	84	51	7	SSW	R-2
Dec	51	34	74	2		24	4	T	21	0	17	14	0	4	0	19	60	79	55	7	SW	R-2
Year	71	51	100	2			43	5	253	1	217	148	42	24	36	76	66	84	53	8	SW	

Notes:
T Indicates "trace"
* For full explanation of (T-H-I) "Temperature Humidity Index," "Wind Chill Factor" and "Storm Intensity," see beginning of Chapter 2.

Average date of first freeze . . . . November 15
" " last " . . . . March 21
" freeze-free period . . . . 239 days
10 inches of snow equal approximately one inch of rain.

RALEIGH, NORTH CAROLINA — Elevation 434 Feet — Table 21

Month	Temperatures Average Max.	Average Min.	Extreme Max.	Extreme Min.	T H I*	Wind Chill Factor*	Precip. Total	Snow	Not even 0.01" precip.	More than ½" of snow	Clear	Cloudy	Thunder-storms	Fog	90° or higher	32° or lower	% of possible sunshine	Rel. Humidity A.M.	P.M.	Wind M.P.H.	Direction	Storm Intensity*
Jan	52	31	77	3			3	3	21	1	16	15	0	3	0	21	56	76	52	9	SW	R, S-2
Feb	54	32	79	8			3	2	18	0	15	13	1	3	0	20	58	72	46	9	SW	R, S-2
March	61	38	89	17			3	1	20	0	17	14	1	2	0	13	65	77	44	10	SW	R, S-2
April	72	47	93	28			4	T	20	0	20	10	4	2	1	2	62	83	46	9	SW	R-2
May	80	56	92	33	73		4	0	21	0	20	11	7	3	1	0	59	85	52	8	SW	T-3
June	86	64	95	43	79		4	0	20	0	22	8	7	2	6	0	60	88	58	7	SW	T-3
July	88	68	97	55	81		6	0	19	0	20	11	11	3	10	0	63	90	61	7	SW	T-4
Aug	87	67	98	46	81		5	0	20	0	21	10	8	3	9	0	61	92	61	7	NE	T-4
Sept	82	60	92	42	78		4	0	21	0	18	12	4	4	2	0	61	92	56	7	NE	R-3
Oct	73	48	89	24			3	0	23	0	19	12	1	4	0	2	64	89	51	7	NNE	R-2
Nov	62	38	81	17			3	T	21	0	20	10	1	3	0	11	62	84	48	8	SW	R-2
Dec	52	31	75	11			3	1	23	0	17	14	0	3	0	19	58	79	52	8	SW	R, S-2
Year	71	48	98	3			44	7	247	1	225	140	45	34	28	88	61	84	52	8	SW	

Average date of first freeze ....November 16
" " last " ....March 24
" freeze-free period ....238 days
10 inches of snow equal approximately one inch of rain.

Notes:

T Indicates "trace"

* For full explanation of (T-H-I) "Temperature Humidity Index," "Wind Chill Factor" and "Storm Intensity," see beginning of Chapter 2.

WILMINGTON, NORTH CAROLINA Elevation 28 Feet Table 22

Month	Temperatures Average Max.	Average Min.	Extreme Max.	Extreme Min.	THI*	Wind Chill Factor*	Precipitation Total	Snow	Not even 0.01" precip.	More than ½" of snow	Clear	Cloudy	Thunder-storms	Fog	90° or higher	32° or lower	% of possible sunshine	A.M.	P.M.	Wind M.P.H.	Direction	Storm Intensity*
Jan	58	37	77	17		24	3	T	22	0	17	14	0	2	0	15	58	78	54	10	N	R-2
Feb	59	38	78	18		24	3	T	18	0	15	13	1	1	0	12	59	74	49	11	NW	R-2
March	65	43	88	25			4	T	20	0	18	13	2	2	0	5	64	78	47	11	SSW	R-2
April	74	51	95	35	76		3	0	22	0	21	9	3	2	1	0	70	79	48	11	SSW	R-2
May	81	60	97	43	80		4	0	22	0	22	9	5	2	3	0	70	81	54	10	SSW	T-3
June	87	68	100	50	82		4	0	21	0	22	8	8	1	7	0	67	83	60	9	SSW	T-3
July	89	71	99	60	82		8	0	19	0	21	10	11	1	13	0	65	84	64	9	SSW	T-4
Aug	88	71	100	57	81		7	0	20	0	22	9	9	1	14	0	65	87	63	8	SW	T-4
Sept	84	66	95	48	79		6	0	20	0	19	11	3	3	2	0	62	87	59	9	N	R-3
Oct	76	55	90	35	71		3	0	23	0	21	10	1	2	0	0	65	85	53	9	N	R-2
Nov	67	44	82	25			3	T	23	0	21	9	1	3	0	3	66	82	48	9	N	R-2
Dec	59	37	81	22		28	3	T	23	0	20	11	0	2	0	13	61	80	54	9	N	R-2
Year	74	54	100	17			51	1.4	253	0	239	126	44	22	38	48	65	81	54	10	SSW	

Notes:

T Indicates "trace"

* For full explanation of (T-H-I) "Temperature Humidity Index;" "Wind Chill Factor" and "Storm Intensity," see beginning of Chapter 2.

Average date of first freeze . . . . .November 24
" " last " . . . . March 8
" freeze-free period . . . . 262 days
10 inches of snow equal approximately one inch of rain.

79

ASHEVILLE, NORTH CAROLINA Elevation 2,140 Feet Table 23

Month	Temperatures Average Max.	Average Min.	Extreme Max.	Extreme Min.	THI*	Wind Chill Factor*	Precipitation in inches Total	Snow	Avg days Not even 0.01" precip.	More than ½" of snow	Clear	Cloudy	Thunder-storms	Fog	90° or higher	32° or lower	% of possible sunshine	Rel. Humidity A.M.	P.M.	Wind M.P.H.	Direction	Storm Intensity*
Jan	48	28	71	−7		16	4	6	20	1	16	15	0	3	0	25	55	86	60	9	NW	R, S-2
Feb	50	28	68	−2		17	4	9	17	1	15	13	0	2	0	22	58	82	56	10	NW	R, S-2
March	56	33	81	14		24	5	4	18	1	17	14	3	3	0	18	72	84	59	10	NW	R, S-2
April	67	42	88	24			4	T	20	0	19	11	4	4	0	4	67	89	54	9	NW	R-2
May	76	50	91	30	69		4	0	20	0	20	11	7	7	0	0	63	93	56	7	NW	T-2
June	82	58	96	35	75		4	0	18	0	21	9	9	9	2	0	64	97	62	6	NW	T-3
July	84	61	94	46	78		6	0	16	0	19	12	13	12	2	0	61	97	66	6	NW	T-4
Aug	83	61	94	43	77		5	0	19	0	21	10	12	14	2	0	60	99	63	6	NW	T-3
Sept	78	54	85	30	74		4	0	21	0	20	10	3	13	0	0	57	99	65	6	NW	R-2
Oct	68	43	83	24			3	0	23	0	22	9	1	7	0	7	63	96	58	7	NW	R-2
Nov	56	33	77	13		26	3	2	22	0	19	11	1	3	0	16	57	89	56	9	NW	R-2
Dec	48	27	68	14		19	4	3	21	1	17	14	0	3	0	23	58	86	58	9	NW	R, S-2
Year	66	43	96	−7			48	24	235	4	226	139	51	79	6	117	62	91	59	8	NW	

Notes:

T Indicates "trace"

* For full explanation of (T-H-I) "Temperature Humidity Index;" "Wind Chill Factor" and "Storm Intensity," see beginning of Chapter 2.

Average date of first freeze .... October 23
"   " " last " .... April 11
"   " freeze-free period .... 195 days
10 inches of snow equal approximately one inch of rain.

CHARLESTON, SOUTH CAROLINA Elevation 40 Feet Table 24

Month	Temperatures Average Max.	Average Min.	Extreme Max.	Extreme Min.	THI*	Wind Chill Factor*	Precipitation in inches Total	Snow	Not even 0.01" precip.	More than ½" of snow	Average number of days Clear	Cloudy	Thunder-storms	Fog	90° or higher	32° or lower	% of possible sunshine	Relative Humidity A.M.	P.M.	Wind M.P.H.	Direction	Storm Intensity*
Jan	61	38	83	11		27	3	T	22	0	16	15	1	4	0	11	56	85	55	9	SW	R-2
Feb	63	40	86	14			3	T	19	0	16	12	1	2	0	8	56	83	52	10	NNE	R-2
March	68	45	90	21			4	T	20	0	16	15	2	2	0	3	66	83	50	10	SSW	R-2
April	77	53	93	29	70		3	0	22	0	19	11	3	2	1	0	67	84	50	10	SSW	R-2
May	84	62	98	36	77		4	0	17	0	20	11	7	2	4	0	68	84	54	9	S	T-3
June	89	69	103	51	81		5	0	20	0	18	12	10	2	11	0	63	86	59	9	S	T-4
July	89	72	101	58	83		8	0	16	0	17	14	14	1	14	0	62	89	64	8	SW	T-4
Aug	89	71	102	58	82		7	0	19	0	20	11	12	1	13	0	63	91	63	8	SW	T-4
Sept	85	66	99	42	80		6	0	20	0	16	14	5	2	4	0	58	91	63	8	NNE	R-3
Oct	77	55	94	27	72		3	0	25	0	19	12	1	3	0	0	65	89	56	8	NNE	R-2
Nov	68	44	88	15			2	T	23	0	20	10	1	4	0	4	64	87	51	8	N	R-2
Dec	61	39	82	8		29	3	T	23	0	17	14	0	3	0	10	62	84	54	9	NNE	R-2
Year	76	55	103	8			49	T	246	0	214	151	57	29	48	37	63	86	56	9	NNE	

Notes:
T Indicates "trace"
* For full explanation of (T-H-I) "Temperature Humidity Index," "Wind Chill Factor" and "Storm Intensity," see beginning of Chapter 2.

Average date of first freeze . . . . December 10
    "   "   " last . . . . February 19
    "   freeze-free period . . . . 294 days
10 inches of snow equal approximately one inch of rain.

81

GREENVILLE, SOUTH CAROLINA — Elevation 957 Feet — Table 25

Month	Average Max	Average Min	Extreme Max	Extreme Min	THI*	Wind Chill Factor*	Precip. Total (in)	Snow (in)	Not even 0.01" precip.	More than ½" of snow	Clear	Cloudy	Thunder-storms	Fog	90° or higher	32° or lower	% of possible sunshine	R.H. A.M.	R.H. P.M.	Wind M.P.H.	Wind Direction	Storm Intensity*
Jan	52	35	76	−6		23	4	3	21	0	18	13	0	3	0	20	54	74	52	7	NE	R-2
Feb	54	36	75	8		24	4	3	17	0	14	14	0	3	0	18	56	72	49	9	NE	R-2
March	62	41	88	17			5	1	19	0	16	15	3	3	0	8	69	73	45	8	SW	R-2
April	71	50	91	32			4	0	20	0	16	14	3	3	0	0	60	78	50	8	SW	R-2
May	79	59	97	38	73		3	0	22	0	17	14	6	2	3	0	60	81	51	7	NE	T-2
June	86	68	98	46	79		3	0	19	0	19	11	7	1	7	0	58	84	56	7	NE	T-2
July	87	71	98	58	80		5	0	18	0	17	14	12	3	10	0	59	87	59	6	WSW	T-3
Aug	86	70	99	52	78		5	0	23	0	21	10	7	3	10	0	64	87	58	6	NE	T-3
Sept	81	64	91	36	76	26	4	0	22	0	17	13	2	2	1	0	63	88	57	6	NE	R-3
Oct	72	52	86	27			3	0	22	0	20	11	1	2	0	1	68	84	51	7	NE	R-2
Nov	62	41	81	17			3	T	23	0	18	12	1	2	0	6	60	78	48	7	SW	R-2
Dec	52	35	71	14			4	T	23	0	17	14	1	5	0	17	54	78	54	7	NE	R-3
Year	71	52	99	−6			46	6	249	0	210	155	41	31	31	71	60	80	52	7	NE	

Average date of first freeze .... November 17
    "   "   "   last   " .... March 23
    "   freeze-free period .... 239 days
10 inches of snow equal approximately one inch of rain.

Notes:
T Indicates "trace"
* For full explanation of (T-H-I) "Temperature Humidity Index," "Wind Chill Factor" and "Storm Intensity," see beginning of Chapter 2.

COLUMBIA, SOUTH CAROLINA — Elevation 213 Feet — Table 26

Month	Temperatures Average Max.	Min.	Temperatures Extreme Max.	Min.	T H I *	Wind Chill Factor*	Precip. Total	Snow	Not even 0.01" precip.	More than ½" of snow	Clear	Cloudy	Thunder-storms	Fog	90° or higher	32° or lower	% of possible sunshine	R.H. A.M.	R.H. P.M.	Wind M.P.H.	Direction	Storm Intensity*
Jan	58	36	76	11		30	3	T	22	0	16	15	1	2	0	18	59	82	54	7	SW	R-2
Feb	61	36	74	9		29	4	T	18	0	15	13	2	2	0	19	60	79	46	8	SW	R-2
March	67	42	87	20			4	T	20	0	17	14	2	1	0	8	65	81	42	8	SW	R-2
April	76	51	92	34	69		4	0	21	0	20	10	4	1	1	0	68	84	46	9	SW	R-2
May	85	60	93	39	76		4	0	23	0	21	10	6	1	4	0	67	89	50	7	SW	T-3
June	92	68	100	52	82		4	0	21	0	22	8	10	2	14	0	64	92	57	7	SW	T-3
July	93	71	99	59	83		6	0	19	0	20	11	13	2	19	0	64	93	58	7	SW	T-4
Aug	91	70	106	53	81		6	0	20	0	24	7	11	2	14	0	67	92	57	6	SW	T-4
Sept	86	65	93	40	80		4	0	22	0	18	12	3	3	2	0	64	94	56	6	NE	R-3
Oct	77	52	89	29	71		2	0	24	0	21	10	1	3	0	2	68	91	51	6	NE	R-2
Nov	67	41	82	21			2	T	23	0	19	11	1	3	0	10	64	88	48	7	SW	R-2
Dec	58	35	79	17		32	4	T	22	0	17	14	0	3	0	18	63	82	52	7	WSW	R-2
Year	76	52	106	9			47	T	255	0	230	135	52	25	54	74	64	87	51	7	SW	

Notes:

T Indicates "trace"

* For full explanation of (T-H-I) "Temperature Humidity Index," "Wind Chill Factor" and "Storm Intensity," see beginning of Chapter 2.

Average date of first freeze .... November 21

     "   " last   " .... March 14

     "   freeze-free period .... 252 days

10 inches of snow equal approximately one inch of rain.

NASHVILLE, TENNESSEE — Elevation 577 Feet — Table 27

Month	Temperatures Average Max.	Average Min.	Extreme Max.	Extreme Min.	THI*	Wind Chill Factor*	Precip. Total	Snow	Not even 0.01" precip.	More than ½" of snow	Clear	Cloudy	Thunder-storms	Fog	90° or higher	32° or lower	% of possible sunshine	Rel. Hum. A.M.	P.M.	Wind M.P.H.	Direction	Storm Intensity*
Jan	49	31	78	−10		19	5	3	19	1	11	20	1	1	0	17	42	84	67	10	S	R, S-3
Feb	52	33	79	−13		22	4	2	17	1	12	16	2	1	0	13	47	80	62	10	S	R, S-3
March	60	40	89	3		28	5	1	19	1	15	16	4	0	0	7	54	78	56	10	S	R, S-3
April	71	49	90	25			4	T	19	0	17	13	5	0	0	1	60	76	51	10	S	R-3
May	79	57	96	36	73		4	T	20	0	18	13	7	1	1	0	65	79	53	8	S	T-3
June	88	66	106	42	80		3	0	21	0	22	8	9	0	9	0	69	79	52	8	S	T-3
July	91	69	107	51	83		4	0	21	0	21	10	10	1	14	0	69	81	53	7	S	T-3
Aug	89	68	105	47	81		3	0	22	0	22	9	7	1	12	0	68	84	54	7	S	T-3
Sept	85	62	105	36	78		3	0	23	0	21	9	4	1	6	0	68	85	52	7	N	R-2
Oct	74	50	94	26	68		3	T	24	0	22	9	1	2	0	1	65	84	51	8	S	R-2
Nov	59	39	85	−1		31	3	T	21	0	18	12	1	1	0	8	55	81	58	9	S	R-2
Dec	50	33	76	−2		21	4	1	20	1	15	16	1	2	0	15	42	82	65	9	S	R, S-3
Year	71	50	107	−13			45	8	246	4	214	151	52	11	42	62	59	81	56	9	S	

Average date of first freeze . . . . November 7
   "   "   last   "   . . . . March 28
   "   freeze-free period   . . . . 224 days
10 inches of snow equal approximately one inch of rain.

Notes:
T Indicates "trace"
* For full explanation of (T-H-I) "Temperature Humidity Index," "Wind Chill Factor" and "Storm Intensity," see beginning of Chapter 2.

MEMPHIS, TENNESSEE — Elevation 262 Feet — Table 28

Month	Temp Avg Max	Temp Avg Min	Temp Extreme Max	Temp Extreme Min	T-H-I*	Wind Chill Factor*	Precip Total	Precip Snow	Days not even 0.01" precip.	Days more than ½" of snow	Clear	Cloudy	Thunder-storms	Fog	90° or higher	32° or lower	% of possible sunshine	Rel. Humidity A.M.	Rel. Humidity P.M.	Wind M.P.H.	Wind Direction	Storm Intensity*
Jan	50	33	79	-8		18	6	2	21	1	12	19	2	1	0	14	44	82	67	12	S	R, S-3
Feb	53	36	80	-11		23	5	1	18	1	13	15	2	1	0	10	51	82	63	12	S	R, S-3
March	62	42	87	12			6	1	20	0	15	16	4	1	0	4	57	79	57	12	S	R, S-3
April	72	52	91	27			5	T	20	0	16	14	5	0	0	0	64	80	54	12	S	T-4
May	80	60	98	38	75		4	0	23	0	18	13	6	0	2	0	68	82	55	9	S	T-3
June	88	68	104	50	80		3	0	22	0	22	8	8	0	12	0	74	84	56	8	S	T-3
July	91	71	106	52	83		3	0	22	0	22	9	8	0	17	0	73	85	56	8	S	T-3
Aug	90	70	105	48	83		3	0	24	0	25	6	7	0	15	0	74	86	54	7	S	T-3
Sept	85	63	103	36	78		3	0	24	0	22	8	3	0	7	0	70	85	52	8	NE	R-2
Oct	76	52	95	25	70		3	T	26	0	22	9	2	0	0	0	69	85	50	9	S	R-2
Nov	61	40	85	9		28	5	T	22	0	18	12	2	1	0	5	58	81	54	10	S	R-3
Dec	53	34	79	2		21	5	1	21	0	15	16	1	1	0	12	45	81	62	11	S	R, S-3
Year	72	52	106	-11			49	5	263	2	220	145	50	6	53	45	64	83	57	10	S	

Notes:

T Indicates "trace"

* For full explanation of (T-H-I) "Temperature Humidity Index," "Wind Chill Factor" and "Storm Intensity," see beginning of Chapter 2.

Average date of first freeze ....November 12
    "   "   " last   " ....March 20
    "   freeze-free period .... 237 days
10 inches of snow equal approximately one inch of rain.

KNOXVILLE, TENNESSEE — Elevation 950 Feet — Table 29

Month	Temperatures Average Max.	Average Min.	Extreme Max.	Extreme Min.	T H I *	Wind Chill Factor*	Precip. Total	Precip. Snow	Not even 0.01" precip.	More than ½" of snow	Clear	Cloudy	Thunder-storms	Fog	90° or higher	32° or lower	% of possible sunshine	Rel. Hum. A.M.	Rel. Hum. P.M.	Wind M.P.H.	Wind Direction	Storm Intensity*
Jan	50	31	77	−16		19	5	3	18	1	12	19	0	2	0	17	42	83	65	7	NE	R, S-3
Feb	53	32	79	−10		21	5	3	16	1	13	15	1	1	0	13	48	81	60	8	NE	R, S-3
March	61	38	88	5		26	5	1	19	0	15	16	3	1	0	8	53	78	54	8	SW	R, S-3
April	71	47	93	23	73		4	T	19	0	17	13	4	0	0	1	60	75	49	8	SW	T-3
May	79	56	96	34	79		4	T	20	0	19	12	6	1	1	0	64	78	51	7	SW	T-3
June	87	65	102	42	81		3	0	20	0	21	9	9	1	7	0	65	81	53	6	SW	T-3
July	89	68	104	52	81		5	0	20	0	20	11	10	1	11	0	64	83	55	6	SW	T-3
Aug	88	66	102	49	80		3	0	21	0	21	10	8	2	9	0	60	87	56	6	NE	T-3
Sept	84	61	103	35	77		3	0	21	0	21	9	4	2	4	0	64	86	52	6	NE	R-2
Oct	73	48	94	24			3	T	23	0	21	10	1	3	0	1	63	86	51	6	NE	R-2
Nov	59	38	84	5		32	3	1	21	0	17	13	1	2	0	8	53	83	56	7	NE	R-2
Dec	50	32	77	−5		23	4	2	20	1	14	17	0	2	0	16	41	84	65	7	NE	R, S-2
Year.	70	48	104	−16			46	9	238	3	211	154	47	18	32	64	56	82	56	7	NE	

Notes:

T Indicates "trace"

* For full explanation of (T-H-I) "Temperature Humidity Index," "Wind Chill Factor" and "Storm Intensity," see beginning of Chapter 2.

Average date of first freeze .... November 6
" " last " .... March 31
" freeze-free period .... 220 days
10 inches of snow equal approximately one inch of rain.

LOUISVILLE, KENTUCKY — Elevation 474 Feet — Table 30

Month	Temperatures Average Max.	Average Min.	Extreme Max.	Extreme Min.	T H I *	Wind Chill Factor*	Precip. Total	Snow	Not even 0.01" precip.	More than ½" of snow	Clear	Cloudy	Thunder-storms	Fog	90° or higher	32° or lower	% of possible sunshine	Rel. Hum. A.M.	P.M.	Wind M.P.H.	Direction	Storm Intensity*
Jan	44	26	79	−20		12	4	4	18	2	10	21	1	2	0	20	41	78	68	10	S	R, S-3
Feb	46	28	78	−19		14	3	4	17	1	12	16	1	1	0	17	47	77	63	10	NE	R, S-3
March	56	35	88	3		22	5	2	18	1	15	16	3	1	0	11	53	75	57	11	NW	R, S-3
April	67	45	91	21			4	T	17	0	15	15	4	0	0	1	57	72	54	10	SW	R-3
May	77	54	98	33	72		4	T	20	0	17	14	6	0	1	0	64	74	53	9	SE	T-3
June	85	63	103	43	78		4	0	19	0	19	11	9	0	7	0	69	76	54	8	S	T-3
July	89	67	107	49	81		3	0	21	0	21	10	8	0	12	0	73	77	52	7	S	T-3
Aug	88	65	105	45	80		3	0	23	0	23	8	7	0	10	0	69	81	54	7	N	T-3
Sept	82	58	104	33	75		3	0	23	0	21	9	4	1	4	0	68	82	53	7	SE	R-2
Oct	71	47	92	23			2	T	24	0	21	10	2	1	0	1	64	81	53	8	SE	R-2
Nov	55	36	83	−1		24	3	1	21	0	16	14	1	1	0	8	51	77	59	9	S	R-2
Dec	46	28	74	−7		16	3	3	21	1	13	18	0	2	0	18	39	78	64	9	S	R, S-2
Year	67	46	107	−20			41	13	242	5	203	162	46	9	34	76	59	77	57	9	S	

Notes:

T Indicates "trace"

* For full explanation of (T-H-I) "Temperature Humidity Index," "Wind Chill Factor" and "Storm Intensity," see beginning of Chapter 2.

Average date of first freeze .... November 7
" " last " .... April 1
" freeze-free period .... 220 days
10 inches of snow equal approximately one inch of rain.

people to be uncomfortable. Nightime readings usually remain in the 70s which is not ideal for sleeping. Traveling north and east, the weather cools off somewhat but still remains quite warm. In the eastern mountains of Tennessee, and also Virginia, afternoon temperatures reach the upper 80s to near 90°, but at night they drop back into the more comfortable 60s. In Delaware and Maryland mid to upper 80s are the rule. The summer months are quite hot in the Carolina area also. Relief can be obtained by going to the coast or west into the mountains. In those areas the temperatures remain in the very pleasant mid 80s on most summer afternoons. Other sections of the Carolinas usually have temperatures in the upper 80s to near 90° in the north and in the low 90s in the south. More comfortable sleeping weather can be found in the mountainous western sections, where nighttime temperatures generally drop to the low 60s. During extreme heat waves, the mercury can soar above the 100° mark in any part of this region, generally accompanied by unpleasantly high humidity.

Summer precipitation throughout the entire area generally falls as afternoon and evening thundershowers. These storms can become quite intense but the possibility of tornadoes is small. There are more thunderstorms and heavier rains at the higher elevations. Two per week would be average in the central and western sections of Tennessee and Kentucky as well as in Maryland, Delaware and Virginia. In the Carolinas, heavy rainfall is also common during the summer, especially in South Carolina. Three very brief thunderstorms per week is about average; tornadoes are possible but not probable. During August the threat of hurricanes becomes evident. Although more strike the Carolinas during the month of September, inhabitants must be on the alert in August. Inland areas experience more thunderstorms, but coastal sections are battered more severely by the very occasional hurricanes. Persons suffering from hay fever will not do too well in these states. Perhaps the only sections with a really good rating are well up in the mountain ranges.

Although this whole wide region is highly popular, it doesn't seem to become overcrowded except in a few spots, and then only during the height of the tourist season.

THE GULF STATES—Area Number 3

See Charts Nos. 20 and 21

(The Florida Panhandle, Georgia, Alabama, Mississippi, Louisiana, Gulfcoast Texas)

This huge 1500-mile crescent bordering the Gulf of Mexico is fast becoming one of America's most popular vacationlands and a haven for retirees.

Actually visitors aren't exactly new to these parts, as many segments of this Gulf country have lived under eight different flags at one time or another. In the ante bellum era much of the center coast was the favorite wintering place of wealthy Northern families. It was a time when this area enjoyed the gracious living and traditional hospitality of the Deep South. It enjoys a more favorable overall climate than the famous French Riviera, with fewer bone-chilling days, but more humid ones.

Although there is more rain in the summer, this area has no real dry season. Spring and early summer herald tornado time, while hurricanes can occur in late summer and fall. Summer rains fall largely as short thunderstorms, but these can be quite heavy. This is also the hot season, often with oppressive humidity. Winters are mild with occasional cold waves, sometimes accompanied by damaging frosts. Snow is a rarity and disappears rapdily.

This is about the largest area of lowlands in the United States made up mostly of coastal sand flats, cypress swamps, and other wet lands and gentle rolling plains. The only high country (1000 ft. elevation) occurs where the tailend of the Appalachians dips down into the northern parts of Georgia and Alabama and just northwest of Austin and San Antonio at the opposite end of the Gulf. Consequently there is little or no mountain weather. All of this adds up to a typical subtropical climate, with all its pluses and minuses.

SUMMER

New Orleans, Mobile, Pensacola and all of the major cities and surroundings in the gulf coast states have a typical subtropical climate. Temperatures throughout the months of June, July and August almost

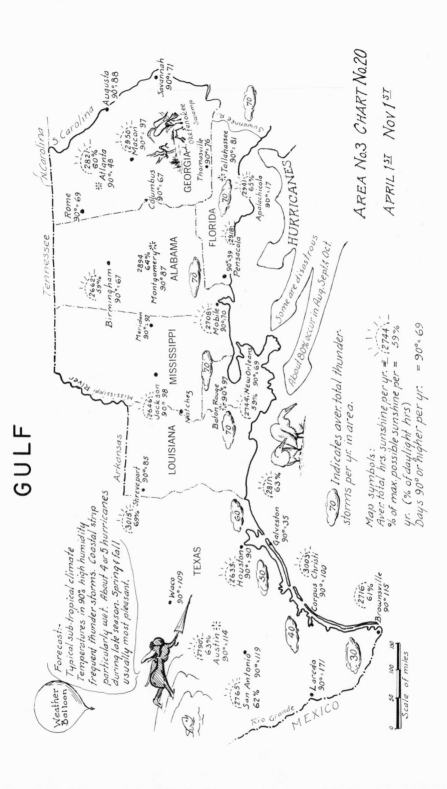

GULF

Weather Balloon

Forecast:-
Typical sub-tropical climate
Temperatures in 90's high humidity
frequent thunder storms. Coastal strip
particularly wet. About 4 or 5 hurricanes
during late season. Spring & fall
usually most pleasant.

AREA No.3 CHART No.20

APRIL 1ST   NOV 1ST

TEXAS

Waco
90°-109

:3015:
69% Shreveport
90°-85

LOUISIANA

Arkansas

Mississippi River

:2646:
Jackson
90°-98

MISSISSIPPI

Natchez

ALABAMA

:2662:
59%
Birmingham
90°-67

Meridan
90°-92

2894
64%
Montgomery
90°-87

:2708:
Mobile
90°-70

70

Baton Rouge
70 90°-97

:2744: New Orleans
59%   90°-69

60

:281in:
63%

:2633:
Houston
90°-90

Galveston
90°-35

50

:2716:
61%
Brownsville
90°-115

Corpus Christi
90°-100

:3003:

40

30

:2765:
62%
San Antonio
90°-119

Rio Grande

MEXICO

:2790:
63%
Austin
90°-114

Laredo
90°-171

Tennessee

Rome
90°-69

N.Carolina

S.Carolina

Augusta
90°-88

Savannah
90°-71

:282:
60%
Atlanta
90°-48

:2950:
Macon
90°-97

GEORGIA

Okefenokee
Swamp

Columbus
90°-67

Thomasville
90°-76

70

Tallahassee
90°-81

:2941:
65%
Apalachicola
90°-17

FLORIDA

70

:29.81in:
90°-39
Pensacola

70

Suwannee R.

70

HURRICANES

Some are disastrous

About 80% occur in Aug. Sept. & Oct.

70 Indicates aver. total thunderstorms per yr. in area.

Map symbols:
Aver. total hrs. sunshine per yr: ≡2744:
% of max. possible sunshine per = 59%
yr. (% of daylight hrs)
Days 90° or higher per yr. = 90°-69

Scale of miles
0   50   100   150

GULF

Forecast: Generally mild & sunny.
Occasional cold snaps dropping into the
teens. Brief showers—light snow—0" to 4".
Pleasant & inviting to tourists
& retirees.

F. D. Powers

Map symbols:
Days 30° or lower per yr. = 41
Days not even 0·01 inches
of precipitation per yr. = 0·01=260
Total inches of snow per yr. = ----2"

AREA No. 3   CHART No. 21
Nov 1ˢᵀ   APRIL 1ˢᵀ

TENNESSEE

North Carolina
6"
4"
2"

Rome
79

GEORGIA

Atlanta
38 0·01=243

Augusta
0·01=253
54

Macon 34
0·01=254

Savannah
30 0·01=251

ALABAMA

Columbus
44

Thomasville
14 0·01=249

Tallahassee
20

Suwannee

MISSISSIPPI

Birmingham
53 0·01=247

Inches of snow per yr.

Montgomery
20
0·01=251

FLORIDA

Apalachicola
4 0·01=257

Meridian
27 0·01=260

Pensacola
9

Mobile
(0·01=240)
18

ARKANSAS

Jackson
36

Natchez

Baton Rouge
18

New Orleans
4 0·01=246

LOUISIANA

Shreveport
31 0·01=266

TEXAS

Galveston
4

Waco
26

Austin
19 Inches of snow 2"-1"

Houston
7
0·01=259

Corpus Christi
5 0·01=296

Laredo
4 0·01=304

San Antonio
0·01=282

Rio Grande

Brownsville
1 0·01=293

MEXICO

2" 1"

4"

4" 2"

1"

6"

Scale of miles
0   50   100   150

always top the 90° mark during the afternoon, occasionally going over 100°. This weather can be extremely uncomfortable when accompanied by a high relative humidity. However, it is frequently relieved by the cooling Gulf breezes particularly along the coast. As an example, New Orleans, ninety miles up from the mouth of the Mississippi, and Mobile back from the coast, can be much less pleasant than Biloxi or Pensacola, which sit right in the path of the refreshing off water breezes. Even in the early morning the relative humidity is about 85%. As the temperatures climb, the afternoon high usually reaches 80–84°, which makes afternoon discomfort almost a certainty for those who have not learned to live with this climate. It's a period for inactivity and enjoying the comforts of air conditioning. As in any sub-tropical atmosphere, a modified siesta for those in a position to indulge is often a very satisfactory solution, since this humid heat is responsible for incidences of heat stroke. Because of the moisture in the air, the evenings and nights remain uncomfortably warm.

Precipitation along the coast is over plentiful, the Alabama segment having the distinction of being one of the wettest sections of the United States. Throughout the gulf states, rain falls almost every other day but principally in the form of rather brief afternoon thunderstorms. Occasionally these storms can become quite severe. Once or twice a year they are accompanied by hail which is sometimes damaging. The frequency of summer rainfall diminishes somewhat as one travels toward the northern sections. Likewise there exists a rather noticeable decrease as one approaches the western Louisiana coast with a steady decline along the Texas coast to Brownsville, which averages about 15″ during the summer period. In the Florida panhandle, Alabama and southeast Mississippi, the traveler can be assured of at least 60%–70% sunshine during June, July and August—while the remainder will be cloudy or rainy. The chances of sunny weather increase in Texas, northwest Mississippi, and Louisiana to 70–80%. A noticeable exception is the Mississippi delta, including New Orleans, where the percentage of possible sunshine is under 60%.

There are a number of spots regarded as being particularly healthy. The Ozone Belt on the north shore of Lake Pontchartrain a few hours north of New Orleans, with its large pine woods, pure artesian spring waters and salubrious atmosphere is one. Another is the single-tax community of Fairhope, perched on the 125-foot bluffs overlooking the bay, across from Mobile. Incidently, this latter is the scene of the unique "jubilees" that may happen one to a half dozen times a year. For some unknown climatic or other reason, and usually occurring about 3:00 in the morning, shrimp, crabs and every manner of gulf citizen, surge into the shallow water and onto the beach where they are gleefully scooped up in baskets by the merry locals.

Those affected by hay fever will be interested in knowing that the entire Gulf coast is rated excellent to good. Exceptions are: New Orleans with very heavy air pollution; Panama City not recommended (although Sunnyside–Panama City Beach rates excellent); Houston and Corpus Christi areas are bad; in fact, just about all of Texas, except the western corner around El Paso, is poor.

Generally the inland parts of all these states are to a greater or lesser degree subject to poor ratings. Birmingham, Alabama is particularly bad.

Table 31	AVERAGE % OF MAXIMUM POSSIBLE SUNSHINE (% of daylight hours)												
Area Number 3 Gulf	Jan	Feb	March	April	May	June	July	Aug	Sept	Oct	Nov	Dec	Year
Georgia													
Atlanta	48	53	57	65	68	68	62	63	65	67	60	47	60
Florida													
Apalachicola	59	62	62	71	77	70	64	63	62	74	66	53	65
Alabama													
Birmingham	43	49	56	63	66	67	62	65	66	67	58	44	59
Montgomery	51	53	61	69	73	72	66	69	69	71	64	48	64
Mississippi													
Vicksburg	46	50	57	64	69	73	69	72	74	71	60	45	64
Louisiana													
New Orleans	49	50	57	63	66	64	58	60	64	70	60	46	59
Shreveport	48	54	58	60	69	78	79	80	79	77	65	60	69
Texas													
Austin	46	50	57	60	62	72	76	79	70	70	57	49	63
Brownsville	44	49	51	57	65	73	78	78	67	70	54	44	61
Galveston	50	50	55	61	69	76	72	71	70	74	62	49	63
San Antonio	48	51	56	58	60	69	74	75	69	67	55	49	62

TALLAHASSEE, FLORIDA

Elevation 64 Feet

Table 32

Month	Temperatures Average Max.	Average Min.	Extreme Max.	Extreme Min.	T H I *	Wind Chill Factor *	Precipitation in inches Total	Snow	Not even 0.01" precip.	More than ½" of snow	Clear	Cloudy	Thunder-storms	Fog	90° or higher	32° or lower	% of possible sunshine	Relative Humidity A.M.	P.M.	Wind M.P.H.	Direction	Storm Intensity *
Jan	65	43	83	15			3	T	23	0	19	12	1	6	0	7	57	87	57	7	NW	R-3
Feb	67	44	84	18			4	T	19	0	16	12	2	5	0	4	60	86	53	7	SSE	R-3
March	72	49	85	23			5	T	22	0	18	13	4	5	0	1	66	86	52	7	NW	T-3
April	79	56	92	32	74		5	0	23	0	21	9	5	3	1	0	71	86	50	7	SE	T-4
May	86	63	102	43	78		4	0	23	0	22	9	8	4	9	0	75	86	49	6	S	T-4
June	91	70	103	56	83		7	0	17	0	19	11	14	2	18	0	71	88	56	5	NW	T-4
July	91	72	100	62	84		8	0	14	0	19	12	19	1	20	0	61	91	64	5	SE	T-4
Aug	90	72	99	63	83		7	0	17	0	21	10	15	1	21	0	61	92	62	5	ENE	T-4
Sept	87	69	98	50	81		6	0	20	0	17	13	7	1	11	0	61	91	61	6	NE	R-3
Oct	81	59	94	33	74		2	0	26	0	24	7	2	2	1	0	70	89	52	6	NE	R-2
Nov	71	47	88	19			2	0	24	0	20	10	1	4	0	2	60	87	53	6	NNW	R-2
Dec	65	43	83	17			3	T	23	0	17	14	1	6	0	6	58	87	58	6	NNW	R-3
Year	79	57	103	15			57	T	251	0	233	132	80	41	80	20	65	88	56	6	NW	

Notes:

T Indicates "trace"

* For full explanation of (T-H-I) "Temperature Humidity Index," "Wind Chill Factor" and "Storm Intensity," see beginning of Chapter 2.

Average date of first freeze .... December 3
" " " last " .... February 26
" freeze-free period .... 280 days
10 inches of snow equal approximately one inch of rain.

94

NEW ORLEANS, LOUISIANA

Elevation 9 Feet — Table 33

Month	Temperatures Average Max.	Average Min.	Extreme Max.	Extreme Min.	T H I*	Wind Chill Factor*	Precip. Total	Snow	Not even 0.01" precip.	More than ½" of snow	Clear	Cloudy	Thunder-storms	Fog	90° or higher	32° or lower	% of possible sunshine	Rel. Hum. A.M.	P.M.	Wind M.P.H.	Direction	Storm Intensity*
Jan	64	48	83	15	73		5	T	21	0	17	14	2	3	0	2	49	85	67	8	S	R-3
Feb	67	50	85	7	78		4	T	19	0	16	12	2	2	0	1	50	84	64	9	S	R-3
March	71	55	90	28	82		7	T	22	0	18	13	4	2	0	0	57	84	60	9	S	T-3
April	78	62	91	38	73		5	0	23	0	20	10	5	1	0	0	63	83	59	8	SSE	T-3
May	84	68	97	50	78		5	0	22	0	23	8	6	0	2	0	68	83	59	8	SSE	T-3
June	89	74	102	58	82		6	0	18	0	23	7	11	0	12	0	65	82	60	7	S	T-4
July	90	76	102	66	83		7	0	16	0	21	10	16	0	17	0	58	84	64	6	S	T-4
Aug	91	76	100	63	84		6	0	17	0	22	9	14	0	17	0	59	85	63	6	SSW	T-3
Sept	87	73	99	54	81		6	0	20	0	22	8	7	0	8	0	64	84	62	7	ENE	T-3
Oct	80	65	94	40	75		4	0	24	0	24	7	2	1	1	0	69	82	58	8	NE	R-3
Nov	70	55	89	29			4	0	23	0	21	9	2	2	0	0	60	83	60	8	NE	R-3
Dec	65	50	84	19			5	T	21	0	17	14	2	3	0	1	46	85	66	8	NE	R-3
Year	78	63	102	7			64	T	246	0	244	121	73	14	57	4	59	84	62	8	S	

Notes:
T Indicates "trace"
* For full explanation of (T-H-I) "Temperature Humidity Index," "Wind Chill Factor" and "Storm Intensity," see beginning of Chapter 2.

Average date of first freeze .... December 12
       "   "   last   "   .... February 13
       "   freeze-free period .... 302 days
10 inches of snow equal approximately one inch of rain.

MOBILE, ALABAMA

Elevation 211 Feet

Table 34

Month	Temperatures				T H I *	Wind Chill Factor*	Precipitation in inches		Average number of days								% of possible sunshine	Relative Humidity		Wind		Storm Intensity*
	Average		Extreme						Not even 0.01" precip.	More than ½" snow	Clear	Cloudy	Thunderstorms	Fog	90° or higher	32° or lower		A.M.	P.M.	M.P.H.	Direction	
	Max.	Min.	Max.	Min.			Total	Snow														
Jan	62	43	84	14			5	T	21	0	15	16	2	5	0	5	46	87	63	12	N	R-3
Feb	65	45	82	11			5	T	18	0	14	14	2	5	0	3	52	87	60	12	S	R-3
March	69	50	90	21			8	T	20	0	17	14	6	3	0	1	56	86	57	12	S	T-3
April	76	57	92	36	71		5	0	22	0	18	12	6	4	0	0	61	87	56	11	S	T-3
May	83	64	100	45	77		5	0	23	0	21	10	8	2	5	0	69	87	55	10	S	T-3
June	89	71	102	56	81		6	0	18	0	22	8	13	0	17	0	71	87	58	9	S	T-4
July	89	72	104	60	82		9	0	13	0	18	13	19	0	19	0	68	90	63	8	S	T-4
Aug	90	72	102	59	83		6	0	18	0	23	8	15	1	21	0	61	91	60	8	NE	T-3
Sept	87	69	98	47	80		6	0	20	0	19	11	8	1	9	0	62	90	61	9	NE	T-3
Oct	79	59	93	32	73		4	0	25	0	23	8	3	2	1	0	65	87	52	9	N	R-3
Nov	69	48	85	22			4	0	23	0	20	10	3	3	0	2	68	84	55	11	N	R-3
Dec	63	44	80	18			5	T	20	0	16	15	2	5	0	4	62	86	62	11	N	R-3
Year	77	58	104	11			68	T	241	0	226	139	87	31	72	15	62	87	59	10	S	

Average date of first freeze .... December 12
    "   "   last .... February 17
    "   freeze-free period .... 298 days
10 inches of snow equal approximately one inch of rain.

Notes:

T Indicates "trace"

* For full explanation of (T-H-I) "Temperature Humidity Index;" "Wind Chill Factor" and "Storm Intensity," see beginning of Chapter 2.

96

MONTGOMERY, ALABAMA   Elevation 198 Feet   Table 35

Month	Temperatures Average Max.	Temperatures Average Min.	Temperatures Extreme Max.	Temperatures Extreme Min.	T H I *	Wind Chill Factor*	Precip. Total	Precip. Snow	Not even 0.01" precip.	More than ½" of snow	Clear	Cloudy	Thunder-storms	Fog	900° or higher	32° or lower	% of possible sunshine	R.H. A.M.	R.H. P.M.	Wind M.P.H.	Wind Direction	Storm Intensity*
Jan	60	39	83	5		30	5	T	20	0	14	17	1	2	0	7	50	84	61	8	NW	R-3
Feb	63	41	84	-5			5	T	18	0	13	15	2	1	0	4	54	82	58	8	S	R-3
March	69	46	90	20			7	T	20	0	16	15	4	0	0	1	61	81	53	8	S	R-3
April	77	53	92	30	73		5	T	21	0	18	12	5	0	0	0	69	80	52	7	S	T-3
May	84	61	99	43	77		3	0	22	0	20	11	6	0	5	0	72	79	51	6	S	T-3
June	91	69	106	48	83		5	0	21	0	20	10	9	0	17	0	72	80	53	6	SW	T-4
July	91	71	107	61	83		6	0	19	0	22	11	11	0	20	0	65	85	58	6	SW	T-4
Aug	91	71	104	58	83		5	0	23	0	22	9	9	0	18	0	69	87	57	5	E	T-3
Sept	88	66	106	45	80		4	0	22	0	19	11	4	0	10	0	68	85	54	6	NE	T-3
Oct	79	54	100	26	73		2	0	25	0	21	*10	1	1	1	0	70	84	50	6	NE	R-2
Nov	67	43	86	13			4	T	23	0	17	13	1	1	0	2	63	84	53	7	NW	R-3
Dec	60	39	83	8		33	5	T	20	0	15	16	1	2	0	6	48	84	61	7	NW	R-3
Year	77	54	107	-5			54	T	254	0	215	150	54	8	71	20	65	83	55	7	S	

Average date of first freeze .... December 3
    "   "   last .... February 27
    "   freeze-free period .... 279 days
10 inches of snow equal approximately one inch of rain.

Notes:
T Indicates "trace"
* For full explanation of (T-H-I) "Temperature Humidity Index;" "Wind Chill Factor" and "Storm Intensity," see beginning of Chapter 2.

97

HOUSTON, TEXAS — Elevation 50 Feet — Table 36

Month	Average Max.	Average Min.	Extreme Max.	Extreme Min.	THI*	Wind Chill Factor*	Precip. Total	Precip. Snow	Days Not even 0.01" precip.	Days More than ½" of snow	Clear	Cloudy	Thunder-storms	Fog	90° or higher	32° or lower	% of possible sunshine	Rel. Hum. A.M.	Rel. Hum. P.M.	Wind M.P.H.	Wind Direction	Storm Intensity*
Jan	64	44	83	17			4	T	21	0	15	16	2	7	0	6	47	84	64	12	NNW	R-2
Feb	66	46	87	25			3	T	18	0	13	15	3	6	0	3	56	84	60	12	SSE	R-2
March	72	51	88	28			3	T	22	0	16	15	2	6	0	1	57	86	59	13	SSE	R-2
April	78	59	93	38	73		3	0	22	0	15	15	4	4	1	0	54	89	63	13	SSE	T-3
May	86	66	94	52	78		4	0	23	0	18	13	7	2	6	0	62	90	61	12	SSE	T-3
June	91	72	100	59	82		4	0	22	0	22	8	7	1	19	0	72	91	61	10	SSE	T-3
July	92	74	101	64	84		4	0	21	0	22	9	10	0	26	0	74	91	59	9	S	T-3
Aug	93	74	106	64	85		4	0	22	0	22	9	9	0	26	0	70	91	59	8	SSE	T-3
Sept	89	69	98	50	82		4	0	21	0	20	10	7	1	15	0	66	88	60	9	SSE	T-3
Oct	82	60	96	38	76		4	0	24	0	22	9	3	4	3	0	73	86	52	10	ESE	R-2
Nov	71	51	88	32			4	0	22	0	18	12	2	5	0	0	60	85	60	11	SSE	R-2
Dec	65	46	82	19			4	T	21	0	15	16	2	6	0	4	50	84	64	11	SSE	R-3
Year	79	59	106	17			46	T	259	0	218	147	59	42	95	13	62	87	60	11	SSE	

Notes:
T Indicates "trace"

* For full explanation of (T-H-I) "Temperature Humidity Index," "Wind Chill Factor" and "Storm Intensity," see beginning of Chapter 2.

Average date of first freeze .... December 11
" " " last " .... February 5
" " freeze-free period .... 309 days
10 inches of snow equal approximately one inch of rain.

BROWNSVILLE, TEXAS — Elevation 19 Feet — Table 37

Month	Temperatures Average Max.	Average Min.	Extreme Max.	Extreme Min.	T H I*	Wind Chill Factor*	Precip. Total	Snow	Not even 0.01" precip.	More than ½" of snow	Clear	Cloudy	Thunderstorms	Fog	90° or higher	32° or lower	% of possible sunshine	R.H. A.M.	R.H. P.M.	Wind M.P.H.	Direction	Storm Intensity*
Jan	71	52	80	32			1	T	24	0	15	16	1	5	0	2	46	89	72	12	SSE	R-2
Feb	73	55	85	32	70		1	T	21	0	13	15	1	5	0	1	49	90	65	13	SSE	R-2
March	77	59	87	35	73		1	T	26	0	15	16	1	3	0	0	50	91	63	14	SE	R-2
April	82	66	93	51	77		2	0	26	0	16	14	2	2	4	0	55	88	61	15	SE	R-2
May	87	71	99	53	80		2	0	27	0	22	9	3	1	11	0	66	88	60	14	SE	T-3
June	91	75	95	66	83		3	0	25	0	24	6	2	0	25	0	72	89	60	13	SE	R-2
July	93	76	96	70	84		2	0	27	0	26	5	2	0	29	0	81	90	55	12	SE	R-2
Aug	93	75	96	63	84		3	0	24	0	25	6	4	0	23	0	76	90	59	11	SE	R-2
Sept	90	73	93	56	83		5	0	20	0	23	7	4	0	10	0	68	90	67	10	SE	R-3
Oct	85	67	91	52	80		4	0	24	0	24	7	2	1	3	0	67	89	61	10	SE	R-2
Nov	77	58	88	41	73		1	0	23	0	17	13	1	3	0	0	53	85	63	11	SSE	R-2
Dec	72	54	87	34			2	T	25	0	15	16	0	5	0	0	46	83	63	11	NNW	R-2
Year	83	65	99	32			27	T	292	0	235	130	24	25	103	3	61	88	62	12	SE	

Average date of first freeze .... December 21
    "   "   last .... February 10
    "   freeze-free period .... 314 days
10 inches of snow equal approximately one inch of rain.

Notes:
T Indicates "trace"
* For full explanation of (T-H-I) "Temperature Humidity Index," "Wind Chill Factor" and "Storm Intensity," see beginning of Chapter 2.

Winter

Winter temperatures are quite pleasant with afternoon readings generally in the 60s and in the 40s at night. On rare occasions winter temperatures soar into the 80s. On the other hand, a few times each season sudden cold waves strike the gulf states, dropping temperatures into the teens.

Precipitation during the winter months occurs about every third day and is in the form of short, light rain showers. At this time the northern sections of these states average slightly more rainfall than along the coast. Snow is rare and, because of mild temperatures, generally disappears in about one day. Although the amount of rainfall in winter is less than in summer, sunny skies are not as prevalent. The summer sunshine is interrupted for only short periods by the heavy afternoon thundershowers. The percentage of sunshine from December through February in southern Florida is 60–70%—highest in the south. The panhandle of Florida, southern Alabama, and the Texas coast from Houston to Brownsville have 50–60% sunshine. Sunny skies prevail 40–50% of the time during this period in Mississippi, Louisiana and Texas, from Houston to the Louisiana border. The remaining daylight hours are cloud covered or rainy.

Table 31 lists the average percentage of sunshine at almost a dozen places along the Gulf. Chart no. 20 shows the total hours of sunshine per year at the same locations.

Spring and Fall

Spring and fall are the most pleasant seasons. Afternoon temperatures usually reach the 70s and 80s—which is quite comfortable for most people. Nights, too, are pleasant with readings usually in the mid 50s creating fine sleeping weather.

However, the spring and fall seasons have disadvantages. Afternoon thunderstorm occurrence is on the increase, averaging about 6–8 days per month during the spring. Some of these spring storms are severe, spawning tornadoes. During autumn, the hurricane threat exists, but again probablities are low. Each gulf coast state has an average of one hurricane in three years.

FLORIDA—Area Number 4

See Charts Nos. 22 and 23

For our purpose, Florida can best be considered in two parts. The panhandle—that portion north of an imaginary line from Jacksonville to Cedar Key, was included in area no. 3, the Gulf country. With its magnolias and moss-hung live oaks, that section is traditionally and climatically much more the Deep South than typical Florida. It even includes the Suwannee (not Swanee) River that Stephen Foster made famous.

This section will concern itself with weather in the 400 miles or so of peninsula south of that point.

If fifty million Frenchmen can't be wrong, perhaps neither can the twenty million Americans who visit Florida each year. This flat hunk of real estate, averaging only 100 feet above sea level, must have something. And indeed it does!

Sunny, mild and generally pleasant, the late fall, winter and spring seasons all classify as very fine to excellent. While Florida summers may be a matter of taste, to many permanent residents it is the favorite season of the year. Actually, more visitors now come in summer than during the winter months.

WINTER

Temperatures: Warm daytime temperatures with generous sunshine and cool evenings create an almost ideal winter vacationland, particularly south of about Daytona Beach.

Miami gets mostly sunny, upper 70 to near 80° days and comfortable, low humidity, with 50 to 60° nights. There is little variation but four or five times each winter there will be a cold snap, usually of about three days duration. However, you can be almost assured of pleasant weather throughout the winter.

While the southern half of the state enjoys about 10% more sunshine than the north, it may be a surprise to many that the east Gold Coast, with the Miami riviera in its center, gets less sunshine than the west Gulf side, which very properly is known as the "sun coast." Actually, the Gold Coast, doesn't fare as well as the overall state, which averages 80 to 100 cloudy days a year as against about 120 in the Miami area.

Table 38 lists the percent of maximum possible sunshine on a

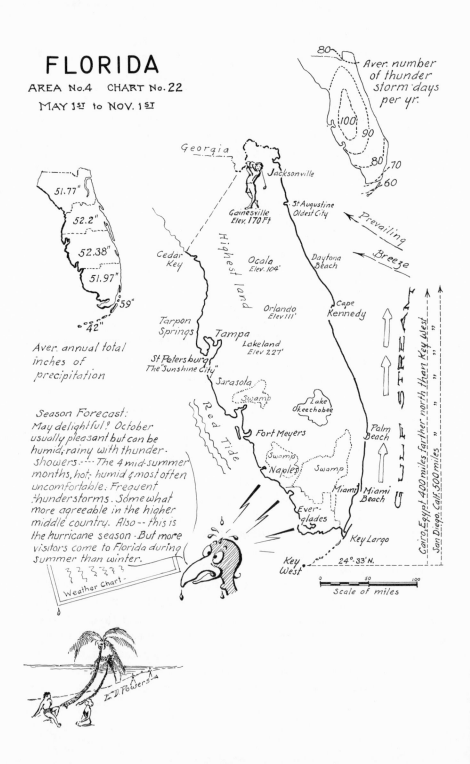

FLORIDA

AREA No.4   CHART No.22

MAY 1ST to NOV. 1ST

Aver. number
of thunder
storm days
per yr.

80
100
90
80 70
60

51.77"
52.2"
52.38"
51.97"
59"
42"

Aver. annual total
inches of
precipitation

Georgia

Jacksonville
Gainesville
Elev. 170 Ft.
St Augustine
Oldest City

Cedar
Key

Highest land

Ocala
Elev. 104'

Daytona
Beach

Prevailing

Breeze

Tarpon
Springs

Tampa
Lakeland
Elev. 227'

Orlando
Elev 111'

Cape
Kennedy

St Petersburg
The "Sunshine City"

Sarasota

Swamp

Red Tide

Lake
Okeechobee

Fort Meyers

Swamp

Naples

Swamp

Miami
Miami
Beach

Ever-
glades

Palm
Beach

GULF STREAM

Key Largo

Key
West

24° 33' N.

Season Forecast:
May delightful! October
usually pleasant but can be
humid; rainy with thundershowers --- The 4 mid-summer
months, hot; humid & most often
uncomfortable. Frequent
thunderstorms. Somewhat
more agreeable in the higher
middle country. Also -- this is
the hurricane season - But more
visitors come to Florida during
summer than winter.

Weather Chart

Cairo, Egypt, 400 miles farther north than Key West
San Diego, Calif. 500 miles

0   50   100
Scale of miles

I.D. Powers

FLORIDA

AREA No. 4   CHART No. 23
NOV 1ST to MAY 1ST

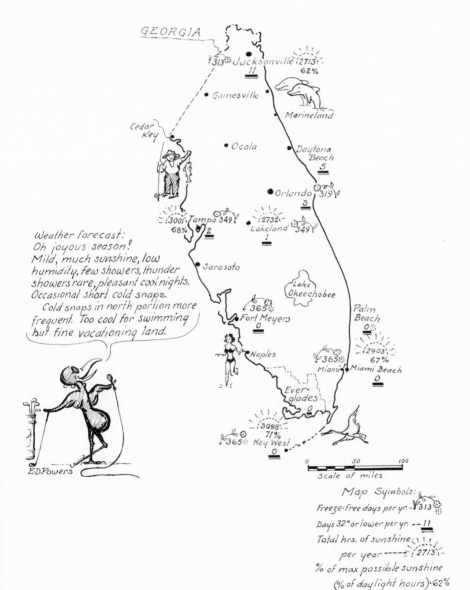

GEORGIA

313 Jacksonville 2713
11   62%

Gainesville

Marineland

Cedar Key

Ocala

Daytona Beach
5

Orlando 319
3

3001 Tampa 349
68%   2

2732
Lakeland 349
1

Sarasota

Lake Okeechobee

365
Fort Meyers
0

Palm Beach
0

Naples

2903
67%

365
Miami   Miami Beach
0

Everglades

3098
71%
365 Key West
0

Weather forecast:
Oh joyous season!
Mild, much sunshine, low
humidity, few showers, thunder
showers rare, pleasant cool nights.
Occasional short cold snaps.
    Cold snaps in north portion more
frequent. Too cool for swimming
but fine vacationing land.

E.D.Powers

0   50   100
Scale of miles

Map Symbols:
Freeze-free days per yr. 313
Days 32° or lower per yr. 11
Total hrs. of sunshine
    per year 2713
% of max possible sunshine
    (% of daylight hours) 62%

monthly basis in key spots. Chart no. 23 also shows the total hours of sunshine per year in five locations from Jacksonville to Key West.

Traveling to the upper northwest corner of Florida we find that Jacksonville doesn't have nearly as good winter conditions as the south. The thermometer seldom registers above the middle 60s, which is too cold for swimming, and drops to the mid 40s at night. Cold waves are much more frequent in northern Florida and the temperature may drop into the 20s.

While Florida is considered a winter resort state, this is much more true of the portion south of Daytona Beach than most of the northern parts. A reasonable gauge is the lower winter rates at many of the motels. The Gulf Stream follows the coast, and is only a few miles off shore at Palm Beach. Somewhat above that point, the coast bends back toward the northwest and the stream heads for Europe where many consider it America's finest export.

The daily winter weather maps show gradually decreasing temperature readings from south to north.

Rainfall is light and occurs perhaps once a week. Winter thunderstorms are rare and skies are mostly sunny. Very occasionally Jacksonville and the northern parts experience a light snow sprinkle and a trace has even been reported as far south as Tampa. Precipitation in southern Florida is in the form of light rain showers.

SPRING AND FALL

Spring is a wonderful time to be in Florida. Ironically most winter visitors leave just before what most agree is the almost perfect period—mid April to mid May. April is second only to November as a dry month. South of Orlando the early spring afternoon temperatures are generally in the upper 70s to near 80s and dip into the 60s and perhaps occasionally the 50s. April warms a bit to the low 80s and in May the thermometer reaches the mid 80s. Shortly thereafter the humidity increases and it becomes less agreeable.

Early fall can be rather uncomfortable with high humidities and 90° temperatures; the night-time mid 70s are not welcomed. It begins to become more agreeable with October's low 80° afternoons and 60s at night. Rain, however, can dampen your spirits a bit.

The robins and first winter visitors begin to appear in November, which is a fine month. Daytime temperatures are in the upper 70s and drop into the mid 50s in the north and 60s in the south during the night. With low humidities and little precipitation, everyone is comfortable.

There are light rainshowers every one or possibly two days a week, in early spring and late fall. Late spring and early fall, however, bring changes. In late May, rainfall and thunderstorm activity increases, particularly in the Miami area, and by June there is a shower or a thunderstorm almost every day.

Early fall is much the same, with showers the rule rather than the exception. This is also the beginning of the hurricane season and although not more than one a year passes over Florida, the idea is very disturbing to many. There have been ten-year periods when there was no hurricane damage at all in Florida. With 1350 miles of shore line from Jacksonville to Pensacola as a target, the chance of a hit in any one spot is very low. Hurricanes are quite unpredictable, but the expectancy is perhaps one in seven to ten years in any one spot in the south and west, and possibly one in ten to twenty years along the northern portion of the east coast.

There is a decrease in shower activity in October but the typical good winter vacation weather begins in November, which is the driest month of the year.

SUMMER

Afternoon temperatures are generally in the hot and humid 90s with still humid night temperatures in the mid 70s. Air conditioning is in general use. There is little change from day to day and there are no real or extended heat waves such as make the headlines in northern areas. It must be remembered that Miami is about 400 miles farther south than Cairo, Egypt and 500 miles south of San Diego, which is at the Mexican border. Like any place with rather extreme weather conditions, the first year is the worst. People soon learn to dress, eat and live in a manner compatible with the climate and in time most suffer relatively little real discomfort. The most comfortable places in Florida during the hottest summer months are right at the shore line or, the uplands, which extend like a backbone down the center of the state from the Georgia line to about Sebring. Many sections of this area are high enough to catch the cooling breezes. Actually, the highest place in these low rolling sand hills is only about 350 feet above sea level.

Rainfall is plentiful, with afternoon thundershowers almost every day. They are usually short and heavy, but most are over in thirty minutes to two hours. Keep in mind that these showers, while an inconvenience, can drop the temperature 5° or more. Although rain averages about eight inches a month, the actual time that it falls is probably less than six to ten per cent of the total day. Florida, which rates as one of the wettest states in the country, has considerable variation in its total precipitation from year to year.

An interesting and unique phenomenon occurs periodically along the Gulf coast, called the "red tide." Millions of minute marine organisms appear in such masses as to usually discourage swimming completely and kill great numbers of fish that either float near the shore or pile up on the hot sand. The stench can become so great that either the fish or residents must vacate.

The cause of the red tide is not known. Some think it's a combination of wind and rain. Others say that the fresh water entering the gulf in flood periods, coupled with certain weather conditions, is the cause. Whatever the reason, climatic or otherwise, the red tide is less welcome than is the coming of the magnolia blossoms. Numerous water spouts (tornadoes at sea) are spotted and about 32 tornadoes per year strike during severe thunderstorms, especially in the St. Petersburg area on the west coast. A good watch is kept for weather disturbances, particularly hurricanes, and ample warning is given of any possible danger.

Although many people who've spent their whole lives in Florida have never had any actual experience or contact with these weather rascals, perhaps you should steer clear of Florida if either thunderstorms or hurricanes happen to be your particular aversion. Florida has the highest rate of thunderstorms in the United States, with the Tampa area

Table 38	AVERAGE % OF MAXIMUM POSSIBLE SUNSHINE (% of daylight hours)												
Area Number 4 Florida	Jan	Feb	March	April	May	June	July	Aug	Sept	Oct	Nov	Dec	Year
Florida													
Jacksonville	58	59	66	71	71	63	62	63	58	58	61	53	62
Key West	68	75	78	78	76	70	69	71	65	65	69	66	71
Miami Beach	66	72	73	73	68	62	65	67	62	62	65	65	67
Tampa	63	67	71	74	75	66	61	64	64	67	67	61	68

JACKSONVILLE, FLORIDA — Elevation 24 Feet — Table 39

Month	Average Max.	Average Min.	Extreme Max.	Extreme Min.	T H I *	Wind Chill Factor*	Precip. Total	Precip. Snow	Not even 0.01" precip.	More than ½" of snow	Clear	Cloudy	Thunder-storms	Fog	90° or higher	32° or lower	% of possible sunshine	Rel. Hum. A.M.	Rel. Hum. P.M.	Wind M.P.H.	Wind Direction	Storm Intensity*
Jan	67	45	85	20			2	T	24	0	19	12	1	5	0	4	59	88	56	9	NW	R-3
Feb	69	47	87	19			3	T	20	0	17	11	1	4	0	3	63	86	52	10	WSW	R-3
March	73	51	90	25			3	T	22	0	19	12	3	3	0	1	63	85	49	10	NW	R-3
April	80	58	93	35	73		4	0	23	0	20	10	4	3	1	0	68	85	47	10	SE	T-3
May	86	65	99	49	77		3	0	23	0	23	8	6	2	10	0	68	82	48	9	WSW	T-4
June	91	71	103	57	82		6	0	19	0	18	12	10	1	18	0	58	84	54	9	SW	T-4
July	92	73	105	65	83		8	0	16	0	19	12	16	1	24	0	59	86	57	8	SW	T-4
Aug	91	73	102	64	83		7	0	17	0	21	10	12	1	22	0	57	89	59	8	SW	T-4
Sept	88	71	100	55	81		8	0	15	0	16	14	6	1	12	0	46	90	61	9	NE	R-3
Oct	80	62	96	38	75		5	0	22	0	18	13	2	3	1	0	52	90	57	9	NE	R-3
Nov	72	51	88	23			2	0	24	0	20	10	1	6	0	1	58	89	55	8	NW	R-2
Dec	67	46	84	17			2	T	23	0	17	14	0	5	0	4	53	88	58	8	NW	R-2
Year	80	59	105	17			53	T	248	0	227	138	61	34	89	11	59	87	54	9	NW	

Notes:
T Indicates "trace"
* For full explanation of (T-H-I) "Temperature Humidity Index," "Wind Chill Factor" and "Storm Intensity," see beginning of Chapter 2.

Average date of first freeze . . . . December 16
"   "   "   last   "   . . . . February 6
"   freeze-free period . . . . 212 days
10 inches of snow equal approximately one inch of rain.

Table 40

ORLANDO, FLORIDA   Elevation 106 Feet

Month	Avg Max	Avg Min	Extreme Max	Extreme Min	T-H-I*	Wind Chill Factor*	Precip Total	Precip Snow	Not even 0.01″ precip	More than ½″ of snow	Clear	Cloudy	Thunder-storms	Fog	90° or higher	32° or lower	% of possible sunshine	R.H. A.M.	R.H. P.M.	Wind M.P.H.	Wind Direction	Storm Intensity*
Jan	71	50	83	29	70		2	T	26	0	23	8	1	6	0	1	62	87	54	9	NNE	R-2
Feb	72	51	88	41	72		2	0	22	0	20	8	1	4	0	0	65	91	54	10	S	R-2
March	76	54	90	40			3	0	24	0	20	11	3	2	1	0	69	89	46	10	S	R-2
April	81	60	94	47	76		3	0	23	0	20	10	5	1	3	0	72	81	40	10	SE	T-3
May	86	66	96	52	79		4	0	23	0	23	8	8	2	17	0	71	88	46	9	SE	T-3
June	89	71	98	62	83		7	0	16	0	20	10	14	1	19	0	70	89	53	8	SW	T-4
July	90	73	100	69	84		8	0	12	0	21	10	20	1	29	0	63	90	51	8	S	T-4
Aug	90	74	99	67	84		7	0	15	0	21	10	16	1	23	0	65	93	57	8	S	T-4
Sept	88	72	95	68	83		7	0	15	0	17	13	9	1	22	0	60	94	56	8	NE	T-3
Oct	83	65	91	49	79		4	0	21	0	20	11	3	2	4	0	60	89	51	9	N	R-2
Nov	76	56	88	49	73		2	0	25	0	22	8	1	3	0	0	69	91	54	9	N	R-1
Dec	72	51	90	29	69		2	0	25	0	20	11	1	4	1	1	62	86	54	9	NNE	R-2
Year	81	62	100	29			51	T	247	0	247	118	81	29	119	2	65	89	51	9	S	

Notes:

T Indicates "trace"

* For full explanation of (T-H-I) "Temperature Humidity Index," "Wind Chill Factor" and "Storm Intensity," see beginning of Chapter 2.

Average date of first freeze .... December 17
  " " " last   " .... January 31
  " freeze-free period .... 319 days
10 inches of snow equal approximately one inch of rain.

MIAMI, FLORIDA Elevation 7 Feet Table 41

Month	Temperatures Average Max.	Average Min.	Extreme Max.	Extreme Min.	T H I *	Wind Chill Factor *	Precip. Total	Precip. Snow	Not even 0.01" precip.	More than ½" of snow	Clear	Cloudy	Thunder-storms	Fog	90° or higher	32° or lower	% of possible sunshine	Rel. Hum. A.M.	Rel. Hum. P.M.	Wind M.P.H.	Wind Direction	Storm Intensity *
Jan	76	58	87	34	73		2	0	25	0	25	6	1	1	0	0	66	86	55	9	NNW	R-2
Feb	77	59	89	32	74		2	0	23	0	22	6	1	1	0	0	72	85	56	10	ESE	R-2
March	80	61	90	40	76		2	0	25	0	23	8	2	1	0	0	73	83	55	10	SE	R-2
April	83	66	93	41	78		4	0	23	0	23	7	3	1	1	0	73	81	56	10	ESE	T-3
May	85	70	94	53	80		6	0	21	0	21	10	7	1	3	0	68	81	59	9	ESE	T-3
June	88	74	98	65	83		7	0	17	0	19	11	12	0	11	0	61	83	63	8	SE	T-4
July	89	75	96	69	84		7	0	14	0	21	10	15	0	17	0	65	84	64	8	SE	T-4
Aug	90	75	98	68	84		7	0	15	0	23	8	16	0	22	0	67	86	63	7	SE	T-4
Sept	88	75	95	69	83		9	0	12	0	18	12	11	0	11	0	62	88	65	8	ESE	T-3
Oct	85	71	92	51	80		8	0	16	0	20	11	7	0	1	0	62	88	64	9	ENE	R-3
Nov	80	65	89	39	76		3	0	22	0	24	6	1	1	0	0	64	87	60	9	N	R-2
Dec	77	59	86	35	74		2	0	24	0	22	9	1	1	0	0	65	86	59	9	N	R-2
Year	83	67	98	32			60	0	237	0	261	104	77	7	66	0	67	85	60	9	ESE	

Notes:

T Indicates "trace"

* For full explanation of (T-H-I) "Temperature Humidity Index;" "Wind Chill Factor" and "Storm Intensity," see beginning of Chapter 2.

Average date of first freeze .... Not Computed
    "   " last   ....   "   "
    "   freeze-free period   ....   "   "
10 inches of snow equal approximately one inch of rain.

TAMPA, FLORIDA — Elevation 19 Feet — Table 42

Month	Average Max.	Average Min.	Extreme Max.	Extreme Min.	T H I*	Wind Chill Factor*	Precip. Total	Precip. Snow	Not even 0.01" precip. More than	More than ½" of snow	Days Clear	Days Cloudy	Thunderstorms	Fog	90° or higher	32° or lower	% of possible sunshine	Rel. Humidity A.M.	Rel. Humidity P.M.	Wind M.P.H.	Wind Direction	Storm Intensity*
Jan	71	51	84	27			2	T	25	0	22	9	1	7	0	1	69	88	57	9	N	R-2
Feb	73	53	87	24			3	T	21	0	20	8	1	3	0	0	69	88	55	9	E	R-2
March	76	56	91	34	71		4	0	23	0	21	10	3	3	0	0	70	86	54	10	S	R-2
April	81	61	93	42	75		3	0	23	0	21	9	4	1	1	0	71	85	51	10	ENE	T-2
May	87	67	96	54	78		3	0	25	0	23	8	6	1	9	0	76	84	52	9	E	T-3
June	89	72	98	62	81		7	0	18	0	21	9	14	0	17	0	67	85	57	8	E	T-4
July	90	73	97	64	83		9	0	13	0	19	12	23	0	20	0	60	88	62	7	E	T-4
Aug	90	74	97	67	83		8	0	14	0	19	12	20	0	21	0	59	90	63	7	ENE	T-4
Sept	89	72	96	59	81		7	0	15	0	18	12	13	0	14	0	61	92	61	8	ENE	T-4
Oct	84	66	94	42	77		3	0	23	0	21	10	4	1	3	0	65	89	57	9	NNE	R-2
Nov	77	57	90	29	71		1	0	25	0	22	8	2	3	0	0	68	88	56	9	NNE	R-2
Dec	73	52	85	26			2	0	25	0	20	11	1	5	0	0	63	87	57	9	N	R-2
Year	82	63	98	24			52	T	250	0	247	118	91	26	85	1	66	88	57	9	E	

Notes:

T Indicates "trace"

* For full explanation of (T-H-I) "Temperature Humidity Index," "Wind Chill Factor" and "Storm Intensity," see beginning of Chapter 2.

Average date of first freeze . . . . December 26
" " " last . . . . January 10
" freeze-free period . . . . 349 days
10 inches of snow equal approximately one inch of rain.

110

KEY WEST, FLORIDA — Elevation 5 Feet — Table 43

Month	Temperatures Average Max.	Average Min.	Extreme Max.	Extreme Min.	THI*	Wind Chill Factor*	Precipitation Total	Snow	Not even 0.01" precip.	More than ½" of snow	Clear	Cloudy	Thunder-storms	Fog	90° or higher	32° or lower	% of possible sunshine	Relative Humidity A.M.	P.M.	Wind M.P.H.	Direction	Storm Intensity*
Jan	74	65	85	49	72		2	0	24	0	26	5	1	0	0	0	69	84	69	13	NE	R-2
Feb	75	66	85	47	74		2	0	23	0	24	4	1	0	0	0	76	82	67	13	SE	R-2
March	77	68	87	55	75		2	0	25	0	26	5	2	0	0	0	73	80	66	13	SE	R-2
April	80	72	88	55	77		2	0	25	0	26	4	2	0	0	0	77	78	64	13	ENE	R-2
May	83	75	91	66	80		3	0	24	0	25	6	4	0	1	0	71	77	65	11	ESE	T-2
June	86	78	94	68	82		4	0	19	0	20	10	8	0	9	0	63	78	67	10	ESE	T-3
July	87	79	95	69	83		4	0	17	0	22	9	13	0	18	0	66	77	66	10	ESE	T-3
Aug	88	79	95	68	83		4	0	16	0	21	10	13	0	20	0	69	78	66	9	ESE	T-3
Sept	87	78	94	70	82		7	0	14	0	21	9	9	0	10	0	60	80	69	11	ESE	R-3
Oct	83	75	92	60	79		6	0	18	0	20	11	5	0	1	0	66	83	69	11	ENE	R-3
Nov	78	71	88	49	75		3	0	22	0	23	7	1	0	0	0	70	84	69	12	ESE	R-2
Dec	75	67	85	46	73		2	0	23	0	23	8	1	0	0	0	67	84	70	12	ENE	R-2
Year	81	73	95	46			40	0	250	0	277	88	57	0	59	0	70	80	67	12	ESE	

Average date of first freeze ....None
" " " last " ....None
" " freeze-free period ....365 days
10 inches of snow equal approximately one inch of rain.

Notes:

T Indicates "trace"

* For full explanation of (T-H-I) "Temperature Humidity Index;" "Wind Chill Factor" and "Storm Intensity," see beginning of Chapter 2.

111

holding the record of about 80 to 90 a year. The frequency reduces to 70 a year in the north and 60 at Key West.

And so it goes. You can consult a student of climatheraphy regarding the benefits of the Florida climate, claimed by many to be one of the healthiest spots in America, or do like twenty million others and enjoy a visit to the Sunny State.

THE GREAT LAKES STATES—Area Number 5

See Chart No. 24

(Western New York, Western Pennsylvania, Ohio, West Virginia, Indiana, Illinois, Wisconsin, Michigan)

This area is an extension of the humid continental climate of the northeastern states, which are characterized by weather extremes and rapid change. Summers are hot but generally very sunny. Thunderstorms are frequent and occasionally severe, but fortunately very brief and usually followed quickly by sunshine. Winters are cold and snowy. As is true in many areas, spring and fall are the choice seasons.

This group of states does not rank at the top of the list of either the retiree or vacationist, being the industrial core of the nation. But there are some outstanding sections that attract visitors in very large numbers throughout most of the year, except winter.

It happens that these places are grouped at the opposite ends of this large area. In the east, there is the pleasant Finger Lakes district of New York state, with its fine orchards, vineyards, meadows, hills and water. Just about every type of outdoor, land and water activity can be enjoyed and many retired persons live here. Parts of Pennsylvania are equally popular, as are West Virginia's mountain resorts and mineral springs. At the west end, the northern peninsula of Michigan and the lake region of Wisconsin are rated most popular by vacationists and sportsmen.

Western New York, Western Pennsylvania, Ohio, and West Virginia

WINTER

January is the coldest month of winter, with afternoon highs usually approaching the freezing point and temperatures at night dropping into the high teens. Cold waves are always a possibility, and extremes have

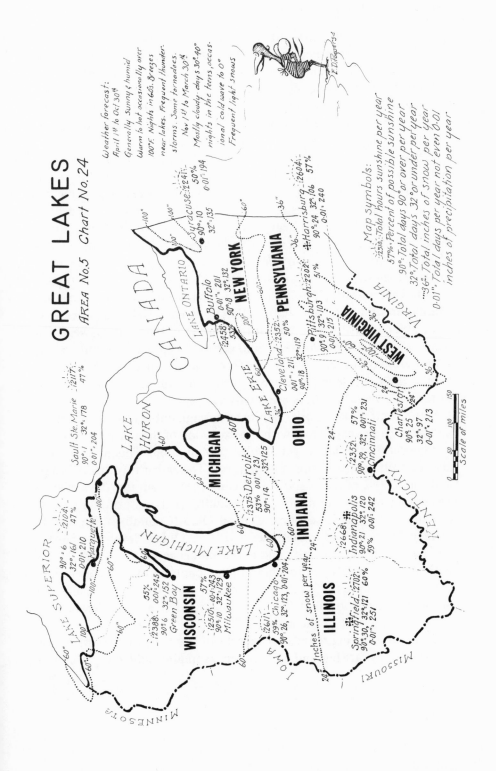

GREAT LAKES
AREA No.5   Chart No.24

Weather forecast:
April 1st to Oct 30th
Generally sunny & humid
Warm to hot occasionally over
100°F. Nights in 60s. Breezes
near lakes. Frequent thunder-
storms. Some tornadoes.
Nov 1st to March 30th
Mostly cloudy days 30°-40°
nights in the teens, occas-
ional cold wave to 0°
Frequent light snows

Map Symbols:
2510: Total hours sunshine per year
57%: Percent of possible sunshine
90°: Total days 90° or over per year
32°: Total days 32° or under per year
-36": Total inches of snow per year
0·01": Total days per year not even 0.01
      inches of precipitation per year

Scale of miles
0   50   100   150

CANADA

LAKE SUPERIOR

Sault Ste Marie :2117:
90°·1   32°·178
0·01"·204

:2104:
47%

Marquette
90°·6
32°·161
0·01"·210

LAKE HURON

LAKE MICHIGAN

WISCONSIN
55%
:2388:   0·01"·243
90°·6   32°·152
Green Bay

57%
:2510: 0·01"·245
90°·10 32°·129
Milwaukee

MICHIGAN

Detroit
:2375: 0·01"·231
53% 90°·14 32°·125

LAKE ONTARIO

Syracuse :2241:
90°·10   50%
32°·135   0·01"·194

Buffalo
:2207:   0·01"·201
90°·8   32°·132

NEW YORK

:2458:
53%

Cleveland :2352:
0·01"·211   50%
90°·18   32°·119

LAKE ERIE

PENNSYLVANIA

Harrisburg :2604:
90°·24   32°·106   57%
0·01"·240

Pittsburgh :2202:
32°·103   51%
90°·25   0·01"·215

OHIO

WEST VIRGINIA

VIRGINIA

INDIANA

Indianapolis
90°·21   32°·120
59%   0·01"·242
:2668:

Inches of snow per year.

ILLINOIS

Springfield :2702:
90°·30, 32°·121   60%
0·01"·251

:2611:
59% Chicago :2510:
90°·26, 32°·123, 0·01"·204

Cincinnati
57%
:2352:   231
90°·29, 32°, 0·01"
0·01"·
:2352:

Charleston
90°·25
32°·97
0·01"·213

:2604:

KENTUCKY

IOWA

MISSOURI

MINNESOTA

E.D.Rogers

dropped the mercury to −30° in all sections. These low temperatures can also be accompanied by strong blustery northwest winds, as any one who has ever fought them on Michigan Avenue in the Windy City will testify. There is no significant warming trend through February, although December is somewhat milder. In December the afternoon temperatures are usually in the mid thirties and drop into the low twenties at night.

When discussing winter precipitation, Ohio, Pennsylvania, and south central New York must be separated from the "snow belt" in north-western New York, along the lee of the Great Lakes. The snow belt which includes such cities as Buffalo, Rochester, Oswego, and Syracuse, receives extremely heavy snowfalls. As cold air from Canada crosses the Great Lakes, it picks up moisture and deposits it in the form of snow across this region. For example, Buffalo averages 23 to 27 inches of snow per month during December, January, and February. Add to that another 25 to 30 inches which fall during November and March, and the total snowfall in Buffalo is over 100 inches per year. Poor Syracuse averages even a little more than that, and Oswego also has tremendous accumulations and can receive huge amounts in very short periods of time; several inches accumulation per hour is common. Clouds are the rule in these areas during the winter. At least six days in ten are overcast; others are partly cloudy.

South central New York (Elmira and Binghamton) and all of central and western Pennsylvania, with the exception of the northwestern corner bordering the lakes, get similar precipitation. This area receives about 8 to 10 inches of precipitation from December through February and about 30% of this is snow. One can expect about two days of snow above one inch accumulation per month. Heavy storms are rarely encountered in these sections. Central Ohio (Columbus) can also be included in the preceding statistics. Northern Ohio (Cleveland), bordering Lake Erie, also comes under the influence of the Great Lakes. Although not nearly approaching the amounts in the snow belt area, the snowfall along the lake is greater than other sections in Ohio. The far southern portion of Ohio (Cincinnati) receives considerably less snow than the rest of the state.

Cloudiness during the winter months in this entire area is the norm, and sunny days are few and far between, averaging only four per month.

SPRING AND FALL

Fall temperatures in this area are warmer than corresponding spring temperatures, due to the influence of the Great Lakes. The lakes freeze over during the winter; therefore surrounding land temperatures stay rather cool till long after the thaw and the lake temperatures slowly recover from the winter cold. For example, in Buffalo, average March temperatures are in the upper thirties during the afternoon and low twenties at night. Afternoon temperatures in November however are in the mid to upper forties, while those at night are in the low thirties. Nearby cities have similar temperature ranges. The month of March is especially susceptible to extremes. Readings in the upper eighties as well as 15 degrees below zero have been recorded in March. April shows a decided warming trend. Afternoon temperatures will usually remain in the low to mid fifties and drop into the low thirties at night. Cities located along the immediate shore of the lakes, such as Oswego, can expect daytime temperatures to remain somewhat cooler. May is usually delightful with daytime temperatures pushing into the comfortable mid and upper sixties and nighttime readings remaining in the mid forties. It is these weather conditions that have enabled the lands adjacent to the lakes, as well as the large Finger Lakes area of New York state, to become such bountiful producers of fruits and berries. Spring is cool enough to prevent too early blossoming and summer long and sunny enough to allow even grapes to mature. Springtime "Peach Blossom Week" can be a glorious sight in these parts. September and October are also very pleasant months. Comfortable seventies are the rule in the afternoons during September, and low fifties can be expected at night. October is cooler with afternoon highs around 60° and nighttime lows around 40° F.

The amount of precipitation is just about the same during spring and fall throughout this entire area. But again springtime is characterized by frequent showery days, whereas in the autumn rain is less frequent though heavier. Although the precipitation pattern does not change significantly in November for most of this area, the sections bordering the lakes do begin to show a definite increase in the number of cloudy days and occurrences of precipitation. For example, Erie and Cincinnati have each about nine days of rain in September; in November, Cincinnati has ten days of precipitation while Erie rises to seventeen. Thunderstorm activity begins and gradually increases through the spring. The sections adjacent to the lakes, however, get fewer.

There will be greater cloudiness in spring than fall, but it should be kept in mind that the number of cloudy days will decrease through the spring as the summer approaches. On the other hand cloudy days are on the increase in autumn.

SUMMER

Summers are sunny, usually hot, and often quite humid. Afternoon June temperatures can be expected to reach into the upper seventies and low eighties. Nighttime temperatures in the mid fifties are common. In July, however, temperatures are generally in the mid eighties during afternoon hours. These high temperatures combined with high relative humidities result in T.H.I.'s in the mid to upper seventies which can be somewhat uncomfortable. Occasional heat waves push temperatures over the 100° mark; then almost everyone is uncomfortable. This is especially true in central and southern Ohio. Areas along the shores of Lake Erie and Lake Ontario, such as Buffalo or Erie, enjoy the cooling effects of the water. This, of course, applies only to the immediate shore line. Mid-town Buffalo, like most large, hemmed-in areas, can be stifling in summer, as afternoon temperatures remain near 80°. The T.H.I. is usually around the comfortable range between 70° and 75°. Rarely do temperatures climb into the nineties. At night one can expect the temperatures to dip to about 60°. The month of August, especially the latter part, is somewhat cooler than July. Heat waves with temperatures over 100° can still occur but this is rare. The low eighties are common in most sections except in central and southern Ohio where mid eighties are the rule. T.H.I.'s are a few points more comfortable during late August.

Summer precipitation generally falls in the form of brief but heavy afternoon thundershowers. In central and southern Ohio one must expect a thundershower every three or four days, especially during July, the hottest month. Areas bordering the Great Lakes experience fewer thundershowers, perhaps one every five or six days. Some thunderstorms may reach great intensity producing damaging hail, high winds and flooding. Even the chance of a tornado exists. In addition to thunderstorms, light showery rains will also occur, about once every two weeks. Long steady rain is infrequent during the summer.

Sunshine is abundant in this area throughout the summer as sunny skies dominate about 70–80% of the daylight hours. Near the lakes this percentage is slightly less, perhaps 65–70%. Table 44 lists the percent of maximum possible sunshine at a number of locations across these states. The detailed map of this area (chart no. 24) shows the total number of hours of sunshine per year at most of these places.

Indiana and Illinois

WINTER

Temperatures in Indiana and Illinois are very similar. Northern areas of these states are colder than the southern parts. For example, January temperatures usually reach into the mid to upper thirties on most afternoons at Indianapolis and Springfield, both located in the central part of each state. Temperatures at night generally drop to near 20°F. Temperatures in these areas can, however, drop well below zero as cold air rushes down from Canada.

The northern sections of these states get somewhat colder weather. Chicago, for example, has afternoon temperatures generally in the low thirties during January with nights in the upper teens. Fort Wayne, Indiana shows a similar range. Fifteen below zero is possible in both of these cities during bitter cold waves. Southern sections of Illinois and Indiana have slightly milder weather. Afternoon temperatures generally reach the 40°F. mark and at night slip down to the mid twenties. The mercury can drop below zero during a cold wave but this is the exception.

Winter precipitation in Illinois and Indiana is generally in the form of rain in the southern sections, such as Evansville, Indiana and points south of Springfield. About 15% of the precipitation in this area falls as snow, resulting in accumulations of about four inches per month during December, January and February. In central Illinois and Indiana, 20% falls as snow, with an average of five inches per month. Northern sections of Illinois and Indiana receive half of their winter precipitation as snow accumulations—about 35 to 40 inches. Seven to ten inches per month is about normal for this area. Most snowfalls are light, but occasionally a severe blizzard cripples the area with heavy snow and gale winds. On January 26, 1967, Chicago experienced a paralyzing snowstorm which dumped 19.8 inches of snow on the suffering public. Wind gusts of more than 60 mph resulted in snow drifts mounting to 12 feet.

Spring and Fall

Fall temperatures are warmer than corresponding spring readings; i.e., September will be milder than May, October milder than April and November milder than March.

In central Indiana and Illinois the weather is quite similar. March temperatures go up to the fifties during the day and drop to near thirty at night. Temperatures, have been recorded in the upper seventies and low eighties, however, as have readings below zero. In April temperatures in the low sixties are the rule with the nighttime mercury dropping into the low forties. May is a most pleasant month in this area. Temperatures in the comfortable seventies are experienced during the afternoon hours, with low fifties at night. Extremes, however, do occur.

The northern sections of this area, such as Chicago and Fort Wayne, are somewhat colder. The March thermometer generally stays in the mid forties during afternoon hours and drops into the mid to upper twenties at night. April temperatures reach about sixty during the day and sink into the upper thirties after dark. May becomes very pleasant with afternoons in the low seventies, and nighttime lows seldom below the upper forties. March is the month of extremes; temperatures do occasionally go below zero but can climb into the eighties.

In all these areas fall is milder than spring by about three to five degrees. Showers are common during spring in this area. Approximately every third day one can expect showers or thunderstorms. During March, however, snow is still a probability. April is usually snow free. As late spring approaches, thunderstorms become more frequent and violent, and are accompanied by high winds, heavy rains and damaging hail. In May and June approximately two thunderstorms per week can be anticipated, and some of these storms spawn tornadoes. Both Indiana and Illinois average 3 to 5 tornadoes per year. Tornadoes most frequently occur in May and June between the hours of 4:00 p.m. and 7:00 p.m. They have, however, been reported for every month of the year and at all hours. The probability that any given area of one square mile will be struck by a tornado in any given year is slight, even in states where these storms are frequent. Warnings are issued by the U.S. Weather Bureau so as to protect citizens as much as possible.

Fall shows a decrease in the occurrence of thunderstorms. Perhaps one a week can be expected during September and this probability is reduced as the season progresses. Showers and prolonged rains continue however, and usually occur about one day in four during the fall. Sunshine is rather abundant during this season.

SUMMER

In Illinois afternoon temperatures reach the mid to upper eighties with T.H.I.'s near 80. Chicago, on Lake Michigan, is somewhat cooler. At night the mid sixties are the rule, which is quite comfortable. Indiana has temperatures very similar to those found in Illinois. In June and August they experience temperatures a few degrees cooler than those in the month of July. It must be remembered, though, that oppressive heat waves can occur during any of the summer months in all of these states and it can be very uncomfortable without air conditioning.

Summer precipitation is mainly in the form of brief but heavy thunderstorms. During June, for example, this happens on one day in three. Fewer storms occur during July and August but still most precipitation is in the form of showers. The threat of a tornado exists throughout the summer. Shower activity decreases slightly as one travels east toward Indiana. Sunshine is abundant through the summer months in this entire region.

Wisconsin and Michigan

WINTER

Temperatures in Michigan vary with latitude. In the north (Sault Ste. Marie and Marquette) temperatures are usually below thirty degrees during the afternoon. At night the mercury drops into the mid teens. During cold waves temperatures of $-30°$ occur. In southern and central Michigan temperatures usually reach the low thirties during the afternoon and drop into the upper teens at night. Again, the thermometer could drop to $-20°$ during arctic-like outbreaks of cold air. Moving further west to Wisconsin, colder weather can be expected. Temperatures in Wisconsin generally climb into the mid to upper twenties during the afternoon and lows at night are in the single numbers. Although temperatures can drop to thirty-below, the coldness is not too penetrating since the air is so dry.

In December Michigan gets 15 to 20 inches of snowfall in the northern and western sections of the state, while the eastern portions get only 8 or 10 inches. These accumulations are the result of many light snowfalls and several heavier snowstorms moving northward from the Mississippi Valley. January is a month of heavy precipitation, mostly snow. Western and northern sections are hardest hit here also with amounts of 15 to 20 inches or more. Eastern parts receive about 10 inches half of which falls as rain. During the winter months in Michigan over one half of the days are cloudy, dreary, and rainy or snowy. February is a drier month. About 10 to 15 inches of snow can be expected (less in eastern portions), but skies remain mostly overcast. A few vigorous snowstorms are possible during February. About 70% of the precipitation is in the form of snow in the western portions and less than half is snow in eastern Michigan. Over 90% is snow in northern sections.

Wisconsin in December averages seven to ten inches, mostly snow. Rain is unusual, in the north, although southern sections still receive rain about forty percent of the time. January is the snowiest month; in the north at this time rain is almost non-existent, but over ten inches of snow must be expected. In the south rain occurs approximately 20% of the time; snow, however, still averages eight to ten inches. Several light snowfalls, combined with a few heavy storms, account for sizable accumulations throughout the state. February snows differ little from the previous month.

Spring and Fall

Temperatures during the month of March are still quite low. In northern sections of Michigan, such as Sault Ste. Marie and Marquette, the mercury reaches the low thirties only during the afternoon. Nighttime temperatures usually drop into the upper teens. In central and southern areas, maximum readings are generally in the low forties, and in the mid twenties at night. April temperatures reach the mid fifties during the afternoons in central and southern sections, with nighttime temperatures dipping into the mid thirties. Northern sections are usually eight to ten degrees cooler. May is also delightful as the weather becomes milder. Mid to upper sixties account for the pleasant afternoons while nights remain in the forties. Extremes can be expected throughout Michigan however, especially during the month of March.

The amount of precipitation during the spring in this area is about the same as the fall; however, in spring it falls mainly as showers and thundershowers, while thundershowers are for the most part absent during fall. Most of the spring thunderstorms can be expected during the month of May, with an average of at least one or more per week.

Table 44	AVERAGE % OF MAXIMUM POSSIBLE SUNSHINE (% of daylight hours)											
Area Number 5 Great Lakes	Jan	Feb	March	April	May	June	July	Aug	Sept	Oct	Nov	Dec	Year
New York													
Binghamton	31	39	41	44	50	56	54	51	47	43	29	26	44
Buffalo	32	41	49	51	59	67	70	67	60	51	31	28	53
Syracuse	31	38	45	50	58	64	67	63	56	47	29	26	50
Pennsylvania													
Harrisburg	43	52	55	57	61	65	68	63	62	58	47	43	57
Pittsburgh	32	39	45	50	57	62	64	61	62	54	39	30	51
Ohio													
Cincinnati	41	46	52	56	62	69	72	68	68	60	46	39	57
Cleveland	29	36	45	52	61	67	71	68	62	54	32	25	50
Columbus	36	44	49	54	63	68	71	68	66	60	44	35	55
Indiana													
Evansville	42	49	55	61	67	73	78	76	73	67	52	42	64
Fort Wayne	38	44	51	55	62	69	74	69	64	58	41	38	57
Indianapolis	41	47	49	55	62	68	74	70	68	64	48	39	59
Michigan													
Detroit	34	42	48	52	58	65	69	66	61	54	35	29	53
Grand Rapids	26	37	48	54	60	66	72	67	58	50	31	22	49
Marquette	31	40	47	52	53	56	63	57	47	38	24	24	47
Sault Ste. Marie	28	44	50	54	54	59	63	58	45	36	21	22	47
Illinois													
Cairo	46	53	59	65	71	77	82	79	75	73	56	46	65
Chicago	44	49	53	56	63	69	73	70	65	61	47	41	59
Springfield	47	51	54	58	64	69	76	72	73	64	53	45	60
Wisconsin													
Green Bay	44	51	55	56	58	64	70	65	58	52	40	40	55
Madison	44	49	52	53	58	64	70	66	60	56	41	38	56
Milwaukee	44	48	53	56	60	65	73	67	62	56	44	39	57

SYRACUSE, NEW YORK — Elevation 424 Feet — Table 45

Month	Average Max.	Average Min.	Extreme Max.	Extreme Min.	THI*	Wind Chill Factor*	Precip. Total	Precip. Snow	Not even 0.01" precip.	More than ½" of snow	Clear	Cloudy	Thunderstorms	Fog	90° or higher	32° or lower	% of possible sunshine	Rel. Hum. A.M.	Rel. Hum. P.M.	Wind M.P.H.	Wind Direction	Storm Intensity*
Jan	34	17	70	−24		0	3	20	11	9	9	22	0	1	0	28	31	76	60	11	NW	S, R-4
Feb	34	17	66	−24		−2	3	19	11	9	8	20	0	0	0	26	38	77	69	11	WNW	S, R-4
March	43	26	84	−16		10	3	15	13	6	12	19	1	1	0	23	45	76	63	11	WNW	R, S-3
April	56	36	90	7		23	3	4	14	1	12	18	2	0	0	10	50	71	56	11	WNW	R-2
May	69	47	92	26			3	T	18	0	16	15	4	0	0	1	58	70	54	9	WNW	R-2
June	79	56	100	34	73		4	0	19	0	19	11	6	0	2	0	64	72	55	8	WNW	T-3
July	84	61	102	44	77		3	0	20	0	21	10	7	0	3	0	67	74	54	8	W	T-3
Aug	81	60	98	42	75		3	0	21	0	20	11	6	0	2	0	63	77	55	8	NW	T-3
Sept	74	53	97	25	69		3	T	19	0	18	12	3	1	1	0	56	80	58	8	S	R-2
Oct	63	42	87	19			3	1	20	0	16	15	1	1	0	3	47	79	59	9	E	R-2
Nov	49	33	81	5		19	3	7	15	3	9	21	0	0	0	15	29	78	66	11	WSW	R, S-2
Dec	37	21	68	−26		5	3	18	12	7	9	22	0	1	0	26	26	78	71	11	WSW	S, R-4
Year	59	39	102	−26			36	83	193	35	169	196	30	5	8	132	50	76	61	10	WNW	

Notes:

T Indicates "trace"

* For full explanation of (T-H-I) "Temperature Humidity Index," "Wind Chill Factor" and "Storm Intensity," see beginning of Chapter 2.

Average date of first freeze .... Mid October
    "   "   last   "   .... Early April
    freeze-free period .... 178 days
10 inches of snow equal approximately one inch of rain.

BUFFALO, NEW YORK   Elevation 693 Feet   Table 46

Month	Average Max.	Average Min.	Extreme Max.	Extreme Min.	T H I*	Wind Chill Factor*	Precip. Total	Precip. Snow	Not even 0.01" precip.	More than ½" of snow	Clear	Cloudy	Thunder-storms	Fog	900 or higher	32° or lower	% of possible sunshine	Rel. Hum. A.M.	Rel. Hum. P.M.	Wind M.P.H.	Wind Direction	Storm Intensity*
Jan	32	19	72	−14		−6	3	19	10	9	8	23	0	1	0	27	32	79	72	17	WSW	S, R-4
Feb	32	17	68	−21		−9	3	17	11	6	8	20	0	1	0	26	41	80	69	16	SW	S, R-4
March	41	25	81	−4	71	1	3	11	15	5	13	18	1	2	0	24	49	79	65	16	SW	R, S-3
April	53	34	87	5	75	17	3	3	15	1	13	17	2	2	0	11	52	77	58	15	SW	R, S-2
May	66	45	94	25	73		2	T	19	0	16	15	4	2	0	1	60	77	56	13	SW	R-2
June	76	55	97	35			3	0	21	0	19	11	5	1	0	0	68	77	56	13	SW	R-2
July	81	60	96	43			2	0	21	0	22	9	6	1	1	0	70	78	53	12	SW	T-2
Aug	79	59	99	43			3	0	21	0	21	10	5	0	1	0	65	82	54	12	SW	T-2
Sept	73	52	98	32			3	T	20	0	18	12	3	1	0	0	60	82	56	13	S	R-2
Oct	60	42	92	24			2	T	20	0	18	13	2	1	0	2	51	82	58	14	S	R-2
Nov	47	33	76	2		16	3	8	14	3	8	22	1	1	0	13	32	80	66	16	S	R, S-3
Dec	35	23	70	−9		−1	3	18	11	6	9	22	0	1	0	24	26	79	71	17	WSW	S, R-4
Year	56	39	99	−21			32	75	198	30	173	192	30	13	2	127	53	79	61	15	SW	

Notes:

T Indicates "trace"

* For full explanation of (T-H-I) "Temperature Humidity Index;" "Wind Chill Factor" and "Storm Intensity," see beginning of Chapter 2.

Average date of first freeze . . . . October 25
  "   "   "   last "   . . . . April 30
  "   "   freeze-free period . . . . 179 days
10 inches of snow equal approximately one inch of rain.

123

ERIE, PENNSYLVANIA — Elevation 732 Feet — Table 47

Month	Average Max	Average Min	Extreme Max	Extreme Min	THI*	Wind Chill Factor*	Total	Snow	Not even 0.01" precip.	More than ½" of snow	Clear	Cloudy	Thunder-storms	Fog	90° or higher	32° or lower	% of possible sunshine	R.H. A.M.	R.H. P.M.	Wind M.P.H.	Wind Direction	Storm Intensity*
Jan	34	20	73	−15		2	2	13	16	8	6	25	0	0	0	27	22	84	79	12	SSW	R, S-3
Feb	34	19	72	−16		1	2	10	14	4	7	21	0	0	0	25	35	85	76	11	WSW	R, S-3
March	43	26	82	−5		8	3	8	17	4	14	17	1	1	0	23	44	81	71	12	SSE	R, S-3
April	54	35	86	7		23	4	3	15	1	13	17	2	1	0	8	51	78	65	10	WSW	R, S-3
May	66	45	91	30			3	T	19	0	15	16	5	1	0	0	58	78	66	8	S	R-2
June	76	55	97	40	73		3	0	21	0	19	11	6	0	1	0	63	80	66	8	S	R-2
July	81	60	98	47	76		3	0	21	0	21	10	7	0	2	0	66	79	64	8	S	T-2
Aug	79	59	96	42	75		2	0	21	0	20	11	5	0	1	0	63	82	63	8	S	T-2
Sept	73	54	99	33	69		3	T	21	0	16	14	4	0	0	0	55	82	63	9	S	R-2
Oct	61	43	89	23			3	1	19	0	17	14	2	0	0	1	44	82	65	9	S	R-2
Nov	48	34	82	6		17	3	8	13	3	8	22	1	0	0	10	24	80	70	11	SSW	R, S-2
Dec	37	24	70	−11		6	2	12	15	7	7	24	0	0	0	23	16	82	75	13	SSW	R, S-3
Year	57	39	99	−16			34	55	212	27	163	202	33	3	4	117	45	81	69	10	S	R, S-3

Average date of first freeze .... November 7
"   "   " last   " .... April 20
"   " freeze-free period .... 200 days
10 inches of snow equal approximately one inch of rain.

Notes:
T Indicates "trace"
* For full explanation of (T-H-I) "Temperature Humidity
Index;" "Wind Chill Factor" and "Storm Intensity," see
beginning of Chapter 2.

PITTSBURGH, PENNSYLVANIA Elevation 1151 Feet Table 48

Month	Average Max.	Average Min.	Extreme Max.	Extreme Min.	THI*	Wind Chill Factor*	Precip. Total	Precip. Snow	Not even 0.01" precip.	More than ½" of snow	Clear	Cloudy	Thunder-storms	Fog	90° or higher	32° or lower	% of possible sunshine	Rel. Hum. A.M.	Rel. Hum. P.M.	Wind M.P.H.	Wind Direction	Storm Intensity*
Jan	37	21	60	−6		4	3	7	15	3	8	23	0	2	0	27	39	78	68	11	WSW	R, S-2
Feb	38	21	69	−7		4	2	4	14	2	9	19	0	3	0	22	42	77	65	11	WSW	R, S-2
March	48	29	72	12		11	3	8	15	2	11	20	2	1	0	21	55	76	56	12	WSW	R, S-2
April	59	38	87	20		25	3	2	16	1	11	19	5	2	0	7	49	76	53	12	WSW	R, S-2
May	71	49	87	32			4	T	19	0	14	17	5	1	0	1	57	74	50	10	WSW	R-2
June	79	58	94	43	73		4	0	19	0	17	13	7	0	2	0	64	79	54	9	WSW	T-3
July	83	62	99	49	77		4	0	20	0	18	13	8	1	4	0	67	81	52	8	WSW	T-3
Aug	81	59	97	46	75		3	0	21	0	20	11	6	1	3	0	57	84	54	8	WSW	T-3
Sept	75	55	97	34	69		3	0	23	0	19	11	4	2	2	0	68	84	52	9	WSW	R-2
Oct	63	43	83	28			2	T	20	0	17	14	1	2	0	3	58	83	56	9	WSW	R-2
Nov	49	33	72	9		19	3	3	18	1	12	18	0	1	0	15	49	78	58	11	WSW	R, S-2
Dec	38	24	66	5		8	3	6	16	3	11	20	0	1	0	23	34	78	66	12	WSW	R, S-2
Year	60	41	99	−7			37	30	216	12	167	198	38	17	11	119	55	79	57	10	WSW	

Notes:

T Indicates "trace"

* For full explanation of (T-H-I) "Temperature Humidity Index," "Wind Chill Factor" and "Storm Intensity," see beginning of Chapter 2.

Average date of first freeze .... October 23
" " last " .... April 20
" freeze-free period .... 187 days
10 inches of snow equal approximately one inch of rain.

CHARLESTON, WEST VIRGINIA — Elevation 950 Feet — Table 49

Month	Temperatures Average Max.	Average Min.	Extreme Max.	Extreme Min.	T H I*	Wind Chill Factor*	Precip. Total	Snow	Not even 0.01" precip.	More than ½" of snow	Clear	Cloudy	Thunder-storms	Fog	900 or higher	32° or lower	% of possible sunshine	R.H. A.M.	R.H. P.M.	Wind M.P.H.	Wind Direction	Storm Intensity*
Jan	46	27	81	−9		17	4	5	14	2	8	23	1	5	0	21	33	78	64	9	S	R, S-2
Feb	49	28	80	−11		18	4	4	14	1	10	18	1	5	0	17	40	78	58	9	WSW	R, S-2
March	57	33	92	2		22	4	3	15	1	11	20	2	3	0	15	44	75	52	10	WSW	R, S-2
April	68	42	96	18			4	T	16	0	13	17	4	3	0	6	49	74	47	9	SW	R-2
May	77	50	98	31	71		4	T	17	0	16	15	8	10	1	0	55	81	50	7	SW	T-3
June	85	60	105	39	78		4	0	19	0	18	12	8	13	6	0	59	86	55	6	SW	T-3
July	87	64	108	46	81		5	0	18	0	18	13	9	15	11	0	62	87	52	6	S	T-3
Aug	86	62	108	47	79		5	0	21	0	19	12	6	19	9	0	58	91	53	5	S	T-3
Sept	81	56	104	33	75		3	0	22	0	18	12	4	14	3	0	58	90	51	6	S	R-2
Oct	71	44	96	18			3	T	21	0	18	13	1	12	0	3	53	88	51	6	S	R-2
Nov	57	35	88	6		29	3	1	18	1	13	17	0	4	0	14	38	79	54	8	SW	R, S-2
Dec	48	29	79	−17		21	3	3	18	1	12	19	1	6	0	21	33	76	60	8	SW	R, S-2
Year	68	44	108	−17			45	17	213	6	174	191	45	109	30	97	48	84	54	7	SW	

Average date of first freeze .... October 28
" " last " .... April 18
" freeze-free period .... 193 days
10 inches of snow equal approximately one inch of rain.

Notes:

T Indicates "trace"

* For full explanation of (T-H-I) "Temperature Humidity Index," "Wind Chill Factor" and "Storm Intensity," see beginning of Chapter 2.

CLEVELAND, OHIO Elevation 787 Feet Table 50

Month	Average Max.	Average Min.	Extreme Max.	Extreme Min.	THI*	Wind Chill Factor*	Precip. Total	Precip. Snow	Not even 0.01" precip.	More than ½" of snow	Clear	Cloudy	Thunder-storms	Fog	90° or higher	32° or lower	% of possible sunshine	Rel. Hum. A.M.	Rel. Hum. P.M.	Wind M.P.H.	Wind Direction	Storm Intensity*
Jan	36	21	73	−9		0	2	11	15	4	8	23	0	2	0	27	27	82	73	13	S	R, S-2
Feb	36	21	69	−8		0	2	10	14	4	9	19	0	2	0	24	34	81	69	13	S	R, S-2
March	45	28	83	−5		18	3	11	15	3	12	19	2	1	0	21	45	80	62	13	WNW	R, S-2
April	57	38	88	19		24	3	3	15	1	13	17	4	2	0	8	50	78	56	12	S	R, S-2
May	70	48	92	29	70		3	0	17	0	16	15	6	1	0	0	57	76	55	11	S	R-2
June	80	58	101	38	74		3	0	19	0	19	11	8	1	5	0	64	78	55	10	S	T-3
July	85	63	103	46	78		3	0	21	0	22	9	7	1	7	0	69	79	52	9	S	T-3
Aug	83	61	102	44	77		3	0	22	0	22	9	5	1	6	0	67	82	53	9	S	T-3
Sept	76	55	101	32	71		3	0	21	0	19	11	3	0	2	0	63	83	53	10	S	R-2
Oct	64	45	90	25		18	2	1	21	0	18	13	2	1	0	1	56	82	55	10	S	R-2
Nov	49	34	82	7			3	7	15	2	10	20	1	1	0	13	33	80	64	13	S	R, S-2
Dec	38	24	69	−9		4	2	11	16	4	9	22	0	1	0	25	28	80	70	13	S	R, S-2
Year	60	41	103	−9			32	53	211	18	177	188	38	14	20	119	49	80	60	11	S	

Notes:

T Indicates "trace"

* For full explanation of (T-H-I) "Temperature Humidity Index;" "Wind Chill Factor" and "Storm Intensity," see beginning of Chapter 2.

Average date of first freeze .... November 2
"   "   " last   " .... April 21
"   freeze-free period .... 195 days
10 inches of snow equal approximately one inch of rain.

COLUMBUS, OHIO — Elevation 814 Feet — Table 51

Month	Average Max.	Average Min.	Extreme Max.	Extreme Min.	T H I*	Wind Chill Factor*	Precip Total	Precip Snow	Not even 0.01" precip.	More than ½" of snow	Clear	Cloudy	Thunder-storms	Fog	90° or higher	32° or lower	% of possible sunshine	Rel. Hum. A.M.	Rel. Hum. P.M.	Wind M.P.H.	Wind Direction	Storm Intensity*
Jan	38	22	74	−20		7	3	7	17	2	9	22	0	1	0	25	36	82	71	10	NW	R, S-2
Feb	40	23	73	−20		8	2	5	16	2	10	18	1	1	0	22	44	81	66	10	NW	R, S-2
March	50	30	85	−1		14	3	3	17	1	12	19	2	1	0	17	49	77	58	11	WNW	R, S-2
April	61	40	90	15		27	3	1	17	0	13	17	4	0	0	5	54	79	55	10	SSW	R-2
May	73	49	96	28	68		4	T	18	0	16	15	8	0	1	0	68	79	55	8	NW	T-3
June	82	59	102	39	76		4	0	18	0	18	12	8	0	5	0	68	81	55	7	SSW	T-3
July	86	63	106	44	78		4	0	21	0	20	11	8	0	8	0	71	83	52	6	SSW	T-3
Aug	84	61	103	42	77		3	0	23	0	22	9	6	0	5	0	68	86	52	6	NNW	T-3
Sept	78	55	100	31	72		3	0	22	0	21	9	4	0	2	0	66	86	50	6	SSW	R-2
Oct	66	43	91	17			2	T	23	0	19	12	1	1	0	2	66	86	53	7	SSW	R-2
Nov	51	33	80	−5			3	2	19	1	13	17	1	1	0	13	44	82	62	10	S	R, S-2
Dec	39	24	70	−14		12	2	4	19	2	11	20	0	1	0	23	35	82	69	9	S	R, S-2
Year	62	42	106	−20			38	22	230	8	184	181	43	6	21	107	55	82	58	8	SSW	

Notes:

T Indicates "trace"

* For full explanation of (T-H-I) "Temperature Humidity Index," "Wind Chill Factor" and "Storm Intensity," see beginning of Chapter 2.

Average date of first freeze .... October 30
" " last " ....April 17
" freeze-free period .... 196 days
10 inches of snow equal approximately one inch of rain.

CINCINNATI, OHIO   Elevation 761 Feet   Table 52

Month	Average Max.	Average Min.	Extreme Max.	Extreme Min.	THI*	Wind Chill Factor*	Precip. Total	Precip. Snow	Not even 0.01" precip.	More than ½" of snow	Clear	Cloudy	Thunder-storms	Fog	90° or higher	32° or lower	% of possible sunshine	R.H. A.M.	R.H. P.M.	Wind M.P.H.	Wind Direction	Storm Intensity*
Jan	41	25	77	−17		3	3	5	18	2	10	21	1	2	0	23	41	82	69	9	SW	R, S-2
Feb	43	27	76	−9		15	3	4	17	2	11	17	1	2	0	20	45	81	64	9	SW	R, S-2
March	53	34	88	6		23	4	3	18	1	12	19	4	1	0	15	53	80	56	9	SW	R, S-2
April	64	43	90	18			4	T	18	0	12	18	4	1	0	4	55	78	52	9	SW	R-3
May	74	53	95	32	69		4	T	19	0	14	17	7	0	1	0	63	80	53	7	SW	T-3
June	83	62	102	40	77		4	0	17	0	17	13	9	0	5	0	69	83	55	7	SW	T-3
July	87	66	109	50	79		4	0	21	0	19	12	10	1	11	0	76	84	52	6	SW	T-3
Aug	85	64	103	43	78		3	0	22	0	21	10	8	1	8	0	69	87	51	5	SW	T-3
Sept	80	58	101	32	74		3	0	21	0	20	10	5	2	4	0	70	87	49	6	SW	R-2
Oct	68	47	92	20			2	T	22	0	19	12	2	3	0	2	59	86	51	6	SW	R-2
Nov	53	36	83	1		28	3	1	20	1	14	16	1	2	0	11	46	81	59	8	SW	R, S-2
Dec	42	28	71	−13		18	3	4	20	2	12	19	0	3	0	22	39	81	66	8	SW	R, S-2
Year	65	42	109	−17			39	17	233	8	181	184	52	18	29	97	57	83	56	7	SW	

Notes:
T Indicates "trace"
* For full explanation of (T-H-I) "Temperature Humidity Index," "Wind Chill Factor" and "Storm Intensity," see beginning of Chapter 2.

Average date of first freeze .... October 25
    "   "   last .... April 15
    "   freeze-free period .... 192 days
10 inches of snow equal approximately one inch of rain.

INDIANAPOLIS, INDIANA — Elevation 793 Feet — Table 53

Month	Average Max.	Average Min.	Extreme Max.	Extreme Min.	THI*	Wind Chill Factor*	Precip. Total	Precip. Snow	Not even 0.01" precip.	More than ½" of snow	Clear	Cloudy	Thunderstorms	Fog	90° or higher	32° or lower	% of possible sunshine	R.H. A.M.	R.H. P.M.	Wind M.P.H.	Wind Direction	Storm Intensity*
Jan	37	21	71	-12		2	3	4	19	1	12	19	1	4	0	27	36	83	72	13	NW	R, S-2
Feb	40	23	72	-19		5	2	4	18	1	12	16	1	3	0	23	51	83	68	12	SE	R, S-2
March	50	30	81	-6		11	4	3	19	2	14	17	3	2	0	20	53	82	61	14	WNW	R, S-3
April	62	40	88	16		25	4	1	18	0	13	17	5	1	0	6	53	80	57	13	W	T-2
May	73	50	93	29			4	T	18	0	16	15	7	1	0	0	61	80	56	11	SW	T-3
June	83	60	102	39	77		4	0	19	0	18	12	8	1	6	0	69	83	58	9	SW	T-3
July	88	64	104	44	80		3	0	22	0	21	10	8	1	8	0	74	84	54	8	SW	T-3
Aug	86	63	100	42	78		4	0	23	0	22	9	6	1	8	0	74	87	54	8	NE	T-3
Sept	79	56	100	28	73		4	0	23	0	20	10	4	1	2	0	70	87	52	9	SW	T-3
Oct	67	44	90	17			3	T	23	0	20	11	2	1	0	3	68	87	53	10	SW	R-2
Nov	51	33	81	-2		18	3	2	20	1	14	16	1	1	0	17	44	84	64	13	W	R, S-2
Dec	39	23	69	-15		5	3	4	21	2	12	19	1	4	0	25	42	84	71	12	SW	R, S-2
Year	63	42	104	-19			40	18	243	7	194	171	47	21	24	121	60	84	60	11	SW	

Notes:
T Indicates "trace"
* For full explanation of (T-H-I) "Temperature Humidity Index," "Wind Chill Factor" and "Storm Intensity," see beginning of Chapter 2.

Average date of first freeze . . . . October 23
    "   "   "   last   " . . . . .April 23
    "   freeze-free period . . . . 182 days
10 inches of snow equal approximately one inch of rain.

130

CHICAGO, ILLINOIS   Elevation 610 Feet   Table 54

Month	Temperatures Average Max.	Average Min.	Extreme Max.	Extreme Min.	T H I*	Wind Chill Factor*	Precip. Total	Precip. Snow	Days Not even 0.01" precip.	More than ½" of snow	Clear	Cloudy	Thunder-storms	Fog	90° or higher	32° or lower	% of possible sunshine	R.H. A.M.	R.H. P.M.	Wind M.P.H.	Wind Direction	Storm Intensity*
Jan	33	19	65	−16		−2	2	9	21	2	12	19	0	2	0	27	44	71	65	11	W	R, S-2
Feb	35	21	59	−9		2	2	8	18	2	13	15	0	2	0	26	47	68	59	12	W	S, R-2
March	44	29	78	8		11	3	7	19	2	14	17	2	1	0	20	51	72	59	12	W	R, S-2
April	57	41	84	26		26	3	1	17	0	14	16	4	1	0	3	52	73	58	12	W	R-2
May	69	51	93	29			4	T	19	0	16	15	5	1	1	0	61	71	51	10	SSW	T-3
June	80	62	96	45	74		4	0	19	0	17	13	7	1	6	0	67	73	53	9	SW	T-3
July	84	67	98	50	78		3	0	22	0	22	9	6	0	7	0	70	76	56	8	SW	T-3
Aug	82	66	98	43	77		3	0	23	0	22	9	5	1	3	0	68	80	56	8	SW	T-3
Sept	75	57	94	40	71		3	0	22	0	21	9	4	1	1	0	64	79	53	9	S	R-2
Oct	63	47	94	27			3	T	24	0	20	11	1	1	0	2	62	76	50	10	S	R-2
Nov	47	33	75	7		15	2	3	20	1	13	17	1	1	0	14	42	76	62	11	SSW	R, S-2
Dec	36	23	63	−10		4	2	10	21	3	14	17	0	2	0	24	41	79	71	11	W	R, S-2
Year	59	43	98	−16			33	38	245	10	198	167	36	14	19	117	57	75	58	10	W	

Notes:

T Indicates "trace"

* For full explanation of (T-H-I) "Temperature Humidity Index," "Wind Chill Factor" and "Storm Intensity," see beginning of Chapter 2.

Average date of first freeze . . . . October 27
   "   "   " last   "   . . . . April 22
   "   freeze-free period . . . . 188 days
10 inches of snow equal approximately one inch of rain.

SPRINGFIELD, ILLINOIS Elevation 587 Feet Table 55

Month	Temperatures Average Max.	Average Min.	Extreme Max.	Extreme Min.	T H I*	Wind Chill Factor*	Precipitation in inches Total	Snow	Average number of days Not even 0.01" precip.	More than ½" of snow	Clear	Cloudy	Thunder-storms	Fog	90° or higher	32° or lower	% of possible sunshine	Relative Humidity A.M.	P.M.	Wind M.P.H.	Direction	Storm Intensity*
Jan	36	21	66	−17		−3	2	5	21	1	12	19	1	2	0	28	45	77	67	13	NW	R, S-2
Feb	41	23	66	−22		0	2	6	19	2	13	15	1	3	0	16	51	78	66	13	NW	R, S-2
March	50	30	82	−12		11	3	4	19	2	15	16	2	2	0	18	51	80	65	14	NW	R, S-2
April	65	42	88	21			4	T	18	0	14	16	6	1	0	5	55	77	55	14	S	T-3
May	76	52	95	28			4	T	20	0	17	14	7	1	2	0	65	79	53	12	SSW	T-4
June	85	63	99	40	77		4	0	19	0	17	13	9	0	9	0	70	80	53	10	SSW	T-4
July	90	66	106	48	79		3	0	22	0	20	11	9	1	8	0	73	84	57	8	SSW	T-3
Aug	87	64	103	43	77		3	0	23	0	21	10	7	1	6	0	75	87	58	8	SSW	T-3
Sept	80	55	99	35	72.		3	0	23	0	21	9	5	1	2	0	73	85	54	9	SSW	T-3
Oct	69	44	91	22			3	T	25	0	21	10	2	1	0	4	67	81	51	11	S	R-2
Nov	52	32	80	−3		14	2	2	21	0	16	14	1	2	0	16	51	81	63	13	S	R, S-2
Dec	40	24	66	−11		0	2	4	22	1	13	18	0	3	0	25	43	80	70	13	S	R, S-2
Year	64	43	106	−22			35	21	252	6	200	165	50	17	27	122	61	81	59	12	S	

Notes:

T Indicates "trace"

* For full explanation of (T-H-I) "Temperature Humidity Index;" "Wind Chill Factor" and "Storm Intensity," see beginning of Chapter 2.

Average date of first freeze . . . . October 30

" " " last " . . . . April 8

" freeze-free-period . . . . 205 days

10 inches of snow equal approximately one inch of rain.

DETROIT, MICHIGAN   Elevation 619 Feet   Table 56

Month	Temperatures Average Max.	Average Min.	Extreme Max.	Extreme Min.	THI*	Wind Chill Factor*	Precip. Total	Snow	Not even 0.01" precip.	More than ½" of snow	Clear	Cloudy	Thunder-storms	Fog	90° or higher	32° or lower	% of possible sunshine	Rel. Hum. A.M.	P.M.	Wind M.P.H.	Direction	Storm Intensity*
Jan	33	21	67	−13		1	2	8	18	3	9	22	0	2	0	28	32	80	70	12	W	R, S-2
Feb	34	20	68	−16		0	2	8	16	3	11	17	1	1	0	26	43	80	67	11	NW	R, S-2
March	42	27	82	−1		9	2	6	18	2	14	17	2	1	0	22	49	79	61	12	NW	R, S-2
April	56	39	87	14		24	3	1	18	0	13	17	3	1	0	8	52	75	53	11	NW	R-2
May	69	49	93	30			4	T	19	0	17	14	5	1	1	0	59	71	51	10	S	T-3
June	79	60	104	38	73		3	0	19	0	18	12	6	0	4	0	65	74	53	9	S	T-3
July	84	65	105	47	77		3	0	22	0	23	8	6	0	6	0	70	75	50	8	S	T-3
Aug	82	64	101	43	75		3	0	23	0	22	9	5	1	4	0	65	81	53	8	N	T-3
Sept	74	56	100	32			2	0	21	0	20	10	3	1	1	0	61	83	54	9	S	R-2
Oct	63	45	92	24			3	T	22	0	19	12	1	1	0	2	56	82	54	10	S	R-2
Nov	47	34	81	5		18	2	3	18	1	11	19	1	1	0	14	35	80	64	11	SW	R, S-2
Dec	36	24	64	−5		7	2	7	18	2	10	21	0	1	0	25	32	80	70	11	SW	R, S-2
Year	58	42	105	−16			31	32	232	11	187	178	32	11	15	125	54	78	58	10	S	

Average date of first freeze . . . . October 21
   "   " last   "   . . . . April 23
   "   freeze-free period   . . . . 180 days
10 inches of snow equal approximately one inch of rain.

Notes:

T Indicates "trace"

* For full explanation of (T-H-I) "Temperature Humidity Index;" "Wind Chill Factor" and "Storm Intensity," see beginning of Chapter 2.

FLINT, MICHIGAN
Elevation 766 Feet
Table 57

Month	Temperatures Average Max.	Average Min.	Extreme Max.	Extreme Min.	THI*	Wind Chill Factor*	Precip. Total	Precip. Snow	Not even 0.01" precip.	More than ½" of snow	Clear	Cloudy	Thunder-storms	Fog	90° or higher	32° or lower	% of possible sunshine	R.H. A.M.	R.H. P.M.	Wind M.P.H.	Wind Direction	Storm Intensity*
Jan	32	17	60	−11		−5	2	10	19	3	10	21	0	2	0	30	30	73	67	12	SW	R, S-2
Feb	33	17	54	−9		−7	2	10	16	3	10	18	0	2	0	26	40	73	63	12	WNW	R, S-2
March	41	24	68	−7		4	2	7	18	2	12	19	1	1	0	25	50	75	62	13	WNW	R, S-2
April	56	35	84	13		19	3	2	18	1	13	17	3	1	0	11	51	75	58	12	WNW	R-2
May	68	45	88	22			3	T	19	0	15	16	5	1	3	2	58	72	51	11	WSW	T-3
June	79	55	94	33	73		3	0	20	0	18	12	6	1	4	0	62	76	51	9	SW	T-3
July	84	59	94	40	76		3	0	22	0	21	10	6	1	2	0	70	81	51	8	SW	T-3
Aug	82	58	97	39	74		3	0	22	0	21	10	6	2	0	0	65	83	55	8	SW	T-3
Sept	73	51	88	31			3	0	21	0	18	12	3	2	0	1	62	88	61	9	S	R-2
Oct	62	41	89	19		28	2	T	22	0	18	13	1	2	0	10	55	80	56	10	SW	R-2
Nov	46	30	72	12		12	2	4	17	1	8	22	1	2	0	16	33	80	67	12	SSW	R, S-2
Dec	35	21	62	−11		0	2	8	19	2	10	21	0	2	0	23	28	79	72	12	SW	R, S-2
Year	58	38	97	−11			30	41	233	12	174	191	33	19	10	145	52	78	60	11	SW	

Notes:

T Indicates "trace"

* For full explanation of (T-H-I) "Temperature Humidity Index," "Wind Chill Factor" and "Storm Intensity," see beginning of Chapter 2.

Average date of first freeze .... October 9
" " last " .... May 8
" freeze-free period .... 155 days
10 inches of snow equal approximately one inch of rain.

GRAND RAPIDS, MICHIGAN — Elevation 681 Feet — Table 58

Month	Average Max.	Average Min.	Extreme Max.	Extreme Min.	THI*	Wind Chill Factor*	Precip. Total	Precip. Snow	Not even 0.01" precip.	More than ½" of snow	Clear	Cloudy	Thunder-storms	Fog	90° or higher	32° or lower	% of possible sunshine	Rel. Hum. A.M.	Rel. Hum. P.M.	Wind M.P.H.	Wind Direction	Storm Intensity*
Jan	31	17	59	−16		−3	2	23	16	6	7	24	0	4	0	30	40	83	72	11	W	S, R-3
Feb	32	16	54	−13		3	2	14	15	4	9	19	0	2	0	27	42	81	67	11	WNW	S, R-3
March	42	24	73	−2		6	2	19	17	4	11	20	2	3	0	27	38	84	67	12	W	R, S-2
April	57	35	81	10			3	3	18	1	12	18	3	3	0	13	50	84	62	12	W	R-2
May	69	45	89	22			3	T	20	0	16	15	4	1	0	2	67	77	50	10	W	R-2
June	79	56	94	37	73		3	0	20	0	18	12	5	1	5	0	71	79	49	9	W	T-3
July	85	60	97	47	77		3	0	21	0	21	10	6	1	8	0	67	82	51	8	W	T-3
Aug	83	58	100	41	75		3	0	23	0	21	10	5	2	3	0	58	89	57	8	W	T-3
Sept	74	50	89	32			3	T	21	0	19	11	4	3	0	1	53	91	60	9	SSW	R-2
Oct	63	40	82	20		30	3	T	22	0	17	14	2	2	0	8	56	86	58	9	SSW	R-2
Nov	46	28	73	10		14	2	11	17	4	8	22	4	3	0	17	31	86	71	11	SSW	R, S-2
Dec	34	20	64	−6		4	2	14	16	6	8	23	0	4	0	25	27	85	77	11	SSW	S, R-3
Year	58	37	100	−16			31	83	226	25	167	198	37	28	16	149	52	84	62	10	W	

Notes:

T Indicates "trace"

* For full explanation of (T-H-I) "Temperature Humidity Index," "Wind Chill Factor" and "Storm Intensity," see beginning of Chapter 2.

Average date of first freeze .... October 27

    " " last " .... April 25

    " freeze-free period .... 185 days

10 inches of snow equal approximately one inch of rain.

135

MILWAUKEE, WISCONSIN — Elevation 672 Feet — Table 59

Month	Average Max.	Average Min.	Extreme Max.	Extreme Min.	T-H-I*	Wind Chill Factor*	Total	Snow	Not even 0.01" precip.	More than ½" of snow	Clear	Cloudy	Thunder-storms	Fog	90° or higher	32° or lower	% of possible sunshine	R.H. A.M.	R.H. P.M.	Wind M.P.H.	Wind Direction	Storm Intensity*
Jan	29	15	62	−24		−9	2	12	21	3	13	18	0	3	0	30	40	76	70	13	WNW	S, R-2
Feb	32	17	60	−19		−6	1	6	19	2	13	15	1	2	0	26	42	76	68	13	WNW	S, R-2
March	41	26	81	−7		4	2	8	20	3	15	16	1	2	0	24	49	78	64	14	WNW	R, S-2
April	53	36	82	13		20	2	1	19	0	14	16	4	3	0	9	52	78	59	14	NNE	R, S-2
May	64	45	90	28			3	T	18	0	17	14	5	3	0	1	56	77	60	13	NNE	T-3
June	75	55	99	33	71		3	0	19	0	18	12	7	3	2	0	61	80	61	11	NNE	T-3
July	81	61	101	45	75		2	0	22	0	22	9	7	1	4	0	70	81	58	10	SW	T-3
Aug	79	61	100	44	74		3	0	22	0	22	9	6	2	4	0	66	83	59	10	SW	T-3
Sept	72	53	98	28			3	T	22	0	20	10	4	1	1	0	61	83	58	11	SW	R-2
Oct	60	42	86	21			2	T	23	0	19	12	1	2	0	3	59	82	57	12	SW	R-2
Nov	45	30	77	−5		10	2	3	20	1	12	18	1	2	0	18	41	80	65	14	WNW	R, S-2
Dec	33	19	63	−12		−3	1	10	21	3	13	18	0	2	0	28	38	77	70	13	WNW	S, R-2
Year	55	38	101	−24			28	39	246	12	198	167	37	26	11	139	54	79	62	12	WNW	

Average date of first freeze .... October 19
"   "   " last .... April 25
"   freeze-free period .... 177 days
10 inches of snow equal approximately one inch of rain.

Notes:
T Indicates "trace"
* For full explanation of (T-H-I) "Temperature Humidity Index;" "Wind Chill Factor" and "Storm Intensity," see beginning of Chapter 2.

136

MADISON, WISCONSIN — Elevation 857 Feet — Table 60

Month	Average Max.	Average Min.	Extreme Max.	Extreme Min.	T H I*	Wind Chill Factor*	Precip. Total	Precip. Snow	Not even 0.01" precip.	More than ½" of snow	Clear	Cloudy	Thunder-storms	Fog	90° or higher	32° or lower	% of possible sunshine	Rel. Hum. A.M.	Rel. Hum. P.M.	Wind M.P.H.	Wind Direction	Storm Intensity*
Jan	28	10	55	−37		−11	1	8	22	3	14	17	0	3	0	30	47	75	67	11	NW	S, R-2
Feb	31	13	58	−25		−7	1	6	20	2	13	15	0	2	0	28	52	78	66	11	WNW	S, R-2
March	42	23	78	−11		5	2	8	20	3	15	16	2	2	0	26	59	79	61	13	NW	R, S-2
April	57	35	90	10		21	2	1	19	0	14	16	4	1	0	12	56	80	53	13	NW	R, S-2
May	69	46	93	27			3	T	20	0	17	14	5	2	0	2	60	77	52	11	WSW	T-3
June	79	56	97	33	73		4	0	19	0	17	13	8	1	4	0	67	80	54	10	S	T-3
July	86	60	102	42	79		3	0	22	0	21	10	8	1	8	0	73	82	52	9	S	T-3
Aug	83	58	101	39	77		3	0	21	0	21	10	5	2	6	0	72	86	53	9	S	T-3
Sept	74	51	99	25	69		4	0	21	0	20	10	4	1	1	1	70	85	51	10	S	T-3
Oct	61	40	87	14		28	2	T	23	0	19	12	1	2	0	6	65	84	52	10	S	R-2
Nov	44	27	74	−11		9	2	4	22	1	12	18	1	2	0	21	42	80	63	12	W	R, S-2
Dec	31	15	60	−21		5	1	10	22	3	13	18	1	3	0	30	41	79	70	11	WNW	S, R-2
Year	57	36	102	−37			30	38	251	12	196	169	39	22	19	156	60	80	58	11	WNW	

Notes:

T Indicates "trace"

* For full explanation of (T-H-I) "Temperature Humidity Index;" "Wind Chill Factor" and "Storm Intensity," see beginning of Chapter 2.

Average date of first freeze .... Mid-October

" " last " .... Late April

" freeze-free period .... 175 days

10 inches of snow equal approximately one inch of rain.

137

March and April, on the other hand, rarely experience more than two or three thundershowers per month. These storms may be severe especially during May, if accompanied by damaging wind and hail, and there is even the chance that one of these thunderstorms may produce a tornado. The significant feature about March precipitation is that much is still in the form of snow.

The northern sections of Wisconsin and Michigan experience significant snowfalls during the month of November, when seven to twelve inches can be expected. Southern portions usually get about three inches during November. Several cities, such as Grand Rapids, which are situated on the eastern side of the lakes, receive about twice as much snow as do the inland cities. Cloudy days are the rule during November in this entire region.

SUMMER

Summer is generally very sunny, warm and at times a bit uncomfortable in most of this area. Central and southern sections usually experience daytime temperatures in the mid to upper 80s. At night the mercury usually drops to near 60°. Readings over 100° are not that uncommon and T.H.I.'s in the mid to upper 70s are the rule. Northern sections, such as Saulte Ste. Marie in Michigan, have temperatures climbing only to the mid seventies on most summer afternoons. Excellent sleeping weather develops as the figures usually read in the low to mid 50s at night. Temperatures of 100° are almost unheard of in this area and the T.H.I.'s generally remain in the low seventies, which is very comfortable.

Thunderstorms comprise the bulk of precipitation during the summer months. Approximately two heavy but very short thunderstorms per week interrupt the sunshine in Wisconsin during summer months. Tornadoes, too, are a possibility and weather warnings should not be ignored. Further east in Michigan thunderstorms occur in the late afternoons and last an hour or less. Sunshine is abundant throughout the entire area. Most days are either clear or partly cloudy but completely cloudy days are the exception.

THE NORTH CENTRAL STATES—Area Number 6

See Chart No. 25

(North Dakota, South Dakota, Iowa, Nebraska, Minnesota)

It's very unlikely that these states will ever become a major retirement center, although several sections draw many enthusiastic vacationists each year. The Black Hills of western South Dakota is a summer playground for much of the vast plain states area. It's a place for camping, riding, fishing or just plain loafing. The free-roving herd of American bison in western South Dakota is maintained at 2500 by disposing of the excess two-year-olds as steaks and roasts. You can buy them alive on the hoof, or as giant "buffalo burgers." Minnesota, which is still 40% forest, boasts the popular 10,000-lake region (actual over 11,000) which remains a largely unspoiled fresh water fishing grounds.

The Badlands and famous Mount Rushmore with its sculptured heads of four Presidents attract many visitors to western South Dakota, as does the autumn pheasant shoot. Otherwise, this is largely agricultural and grazing country, although manufacturing generates three times as great revenue as all the crops and herds combined.

This whole area is characterized by low humidity, severe winters, hot sunny summers, and frequent changes in the weather. Summer thunderstorms are common but occur with greater frequency in Iowa and Nebraska than in the Dakotas. Several tornadoes occur nearly every year and heavy snowfalls are not infrequent.

WINTER

The Dakotas are extremely cold in winter. North Dakota maximum temperatures range from the mid teens in the north to near 20° in southern sections. At night temperatures below zero are common throughout the state. Cold air from Canada drops the temperatures to thirty below zero several times during a winter. Temperatures in the mid 20s are common in the northern parts of South Dakota while above freezing figures are more usual in the southern portions. At night, temperatures in the single figures are experienced in northern sections; while temperatures nearer ten degrees are more common in southern areas, severe cold waves can drop them to −20 or lower. Because the air is usually dry in the Dakotas, these extremely low temperatures are

139

not as uncomfortable as one might think—unless a strong wind accompanies this cold weather.

Minnesota temperatures are even lower. In northern sections, such as Duluth and International Falls (one of the coldest spots in the United States) temperatures remain mostly in the teens in the afternoon, but below zero readings at night are the rule. Special heating facilities are necessary for cars as temperatures drop to −30° and lower. Although again the dry air lessens the impact of the extreme cold, much protective clothing must be worn. In southern sections, Minneapolis, for example, maximum temperatures are usually in the low 20s and single numbers are common at night. Temperatures of −20° or lower are experienced when arctic air plunges southward from Canada.

Iowa and Nebraska residents rarely see the mercury rise above the 30s during January. Sioux City, Iowa, for example, struggles hard to reach the thirty-degree mark during the afternoon. Night time temperatures drop into the low to mid teens. But 15 to 20 below zero readings are experienced with the occasional cold waves which strike this area. December and February are slightly warmer as can be seen from the charts.

Eastern sections of the Dakotas usually receive about 6″ of snow per month while western portions get about 4″. Light snowfalls can be expected at least once a week. Severe storms move northward from the Texas area several times during a winter and produce blizzard conditions.

Minnesota rarely experiences any form of precipitation other than snow throughout the entire winter. Snowfalls generally average nine to ten inches per month. The low temperatures prevent melting and the ground always has a substantial snow cover. Blizzard conditions exist occasionally in this area with cold temperatures, falling snow and high winds. Lack of moisture, because of the distance from a large body of water, prevents this state from receiving very sizable accumulations of snow.

In Nebraska there is more snow in the eastern sections, but rain can and does occasionally fall. In the western part of the state, lighter precipitation, mostly in the form of snow, usually falls in association with light to moderate storms. As in the Dakotas, occasional severe winter snowstorms strike, paralyzing the area for several days. In Iowa seven of ten storms during winter bring snow rather than rain; eight to ten inches of snow per month is normal. Iowa, too, falls prey to the

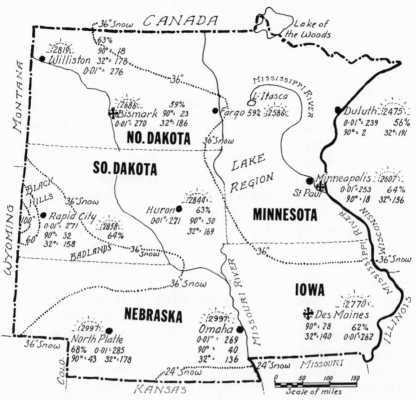

CANADA

Lake of the Woods

MONTANA

36" Snow

63%
90°: 18
:2819:
● Williston 32°: 178
0·01": 276

36"

MISSISSIPPI RIVER

L. Itasca

:2686:   59%
✠ Bismark 90°: 23
0·01": 270 32°: 186

Fargo 59% :2586:

NO. DAKOTA

Duluth :2475:
0·01": 239   56%
90°: 2   32°: 191

36" Snow

SO. DAKOTA

LAKE

REGION

Minneapolis :2607:
0·01": 253   64%
St Paul   90°: 18 32°: 156

WYOMING

BLACK
HILLS

36" Snow

100°
● Rapid City
0·01": 271
60°   :2858:
32°: 158   64%

90°: 32

BADLANDS   36" Snow

:2844:
Huron   63%
0·01": 271 90°: 30
32°: 169

MINNESOTA

MISSISSIPPI RIVER

WISCONSIN

36" Snow

36" Snow

MISSOURI RIVER

36"

IOWA

:2770:
✠ Des Moines
90°: 28   62%
32°: 140 0·01": 262

ILLINOIS

NEBRASKA

:2997:
Omaha
North Platte   0·01": 269
68% 0·01": 285   90°: 40
90°: 43 32°: 178   32°: 136

:2997:

36" Snow

COLO.

24" Snow   24" Snow   MISSOURI

KANSAS

Scale of miles
0   50   100   150

E.D.Powers

Map symbols:
:2997: = Total hours sunshine per year
68% = Percent of possible sunshine
90° = Total days 90° or over per year
32° = Total days 32° or under per year
36" Total inches of snow per year
0·01" = Total days per year not even
    0·01 inches of precipitation per year

Weather forecast: May through Sept, Hot, sunny
pleasant days (80°- 90°), nights 50° to 60° occasional 100° heat waves, severe thunder storms &
tornadoes especially in Nebraska & Iowa.

Forecast: October through April
Mostly cloudy frequent light
snows but occasional heavy storms
& cold waves well below zero. Days
20° to 30° · Nights in teens to
    zero

NORTH CENTRAL

Area No. 6
Chart No. 25

same occasional blizzards which assault the Dakotas and Nebraska. Sioux City receives almost all of its precipitation as snow.

SPRING AND FALL

Spring temperatures in the Dakotas are about 4 or 5 degrees cooler than the corresponding fall figures. During March maximum temperatures can be expected to reach the low to mid 30s in North Dakota and 40° in South Dakota At night, they usually drop into the mid teens in North Dakota and around 20° in South Dakota. Spring is a time of rapid change. March temperatures can go into the 70s but they may also sink to 30 below zero in North Dakota. South Dakota experiences temperatures near 80° but the mercury may also zoom down to zero during March. Again, the dryness of the air takes some of the bite out of the cold. April sees readings in the low to mid 50s in North Dakota during daytime hours and near 30° at night. In South Dakota afternoon highs will be in the upper 50s and the mid 30s at night. Extremes can still occur in April—reaching the upper 80s or dropping into single numbers. May readings are almost the same in North and South Dakota: upper 60s to near 70 are average for the afternoon with temperatures in the mid 40s at night. May can be very warm, with an occasional 90° but during cold outbreaks the low twenties are also possible. The important fact to remember is that these weather changes are not only very great, but can also be very rapid.

Minnesota is an exceptionally cold area. During the month of March temperatures reach no higher than the low 30s in the northern part of the state and the mid 30s in southern sections. At night single-number temperatures are not unusual in northern sections, such as International Falls, while it is usually ten degrees warmer in southern parts, such as Minneapolis. As in so many other areas, March is an extremely changeable month in Minnesota with the thermometer occasionally registering into the 70s or dropping to −20° during cold waves. In April temperatures hover in the mid to upper 40s during the daytime in northern sections, and in the mid 50s in the southern sections. Nights will be in the 20s in the north but remain in the mid 30s in the south. Highs can be in the 90s during April or the mercury may drop into the low teens. May readings rise to the high 60s to near 70° in southern sections during daytime hours and fall into the mid 40s at night. In

142

northern sections you can expect day temperatures to register in the low 60s and at night in the upper 30s. May is much more stable than March and April. Autumn temperatures are 5–6 degrees warmer than corresponding spring figures. For example, October is warmer than April, September warmer than May, and November warmer than March.

Nebraska and Iowa are also noted for weather extremes at this time of year. March can experience temperatures in the upper 80s or in the 10 to 15 below zero range. In general March afternoon readings are in the mid to upper 40s and drop down to the upper 20s at night. April maximums are in the low to mid 60s; with minimums around 40°. In May, the day figures approach the mid 70s while at night are usually in the low 50s. Fall temperatures are somewhat warmer.

Precipitation is on the upswing during the month of March, especially in South Dakota. About one half is in the form of snow in eastern South Dakota which accumulates up to 7″. In western South Dakota perhaps ¾ of the precipitation is snow piling up about 8″ during March. North Dakota remains rather dry through March and although most of the precipitation falls as snow only about 6″ can be expected. April is still a snow month in the Dakotas, with South Dakota receiving about 6″ and North Dakota about 3″. May rarely experiences much snow. Precipitation generally falls as light rain showers or occasional thunderstorms. Expect rain showers about every third day in North Dakota; a little more often in South Dakota.

Fall is usually quite dry in the Dakotas. Thunderstorms become infrequent and during the month of September most of the precipitation falls as rain which is experienced only about once a week and then is usually light. October, although still dry, finds snow in the air. South Dakota receives 2″ of snow compared to 1″ for North Dakota; you can expect precipitation to occur about once a week. Skies are usually cloudy in North Dakota during November but precipitation is rather light. About 6″ of snow falls in North Dakota while 4″ accumulates in western South Dakota and 3″ in eastern South Dakota. Much of the precipitation in eastern South Dakota still falls as rain during November. While spring precipitation is largely in the form of showers and thundershowers, in March it is still mostly snow. Northern Minnesota receives most of its precipitation in the form of snow in March, which accumulates to a depth of about 12″. Southern sections receive somewhat less. Occasional thundershowers occur in April and by May one thunderstorm per week is average.

The northern sections of Minnesota experience significant snowfalls during the month of November with an accumulation of 7 to 12 inches, while the southern portions usually receive about 3 inches. Cloudy days are the rule during this month.

March precipitation in Iowa and Nebraska is frequently in the form of snow. Western sections of Nebraska will get from 5 to 7" of snow during March with northern sections receiving more than southern areas. Heavy spring storms are possible in March, but in general storms are light to moderate. April, however, introduces some violent weather, anticipating the cyclone season of late spring and early summer. Great extremes in April temperatures are instrumental in forming vigorous thunderstorms. About one such thunderstorm can be expected each week, and although they only deposit approximately 2 to 4" of rain in the area, they are nevertheless violent in nature. May is characterized by more frequent (two per week) heavy thunderstorms, and tornadoes become a distinct possibility. These storms, although severe, usually are brief and most likely during the afternoon hours. Damaging winds and hail also occur frequently and weather bureau bulletins should be followed closely.

Autumn thunderstorms are neither as frequent nor as severe. Nebraska and Iowa weather is generally pleasant through September and October. November begins the snow season and some 3" will fall during that month.

SUMMER

Although the air is very dry in the Dakotas, some discomfort may be experienced because of the high afternoon temperatures. T.H.I.'s are generally in the upper 70s during the afternoon but the dry air causes temperatures to fall rapidly at night into the mid to upper 50s. Readings well over 100° have been recorded when warm air flows northward from the Gulf region. These heat waves can be expected two or three times a summer.

This whole area is blessed with a great abundance of sunshine throughout the entire summer. Table 61 shows the percent of maximum possible sunshine in a scattering of places across these states. The detailed map (chart no. 25) shows the number of total hours of sunshine per year at these locations.

Nebraska summers are hot and sometimes quite humid. Afternoon temperatures generally reach the low nineties and this causes some people to be uncomfortable when combined with moderately high humidities. At night however they drop into the moderately pleasant upper 60s. During heat waves, temperatures can soar well over the 100° mark. June and August are slightly cooler than July.

This is pleasant summer country. Northern sections, such as Duluth, have temperatures climbing only to the mid 70s on most summer afternoons and at nighttime slide into the low to mid 50s. 100° in this area is almost unheard of and T.H.I.'s generally remain in the comfortable low 70s. Southern Minnesota is warmer with the afternoon thermometer reaching the mid to upper 80s. At night temperatures drop to the low 60s but an occasional heat wave can bring the daytime

Table 61	AVERAGE % OF MAXIMUM POSSIBLE SUNSHINE (% of daylight hours)											
Area Number 6 North Central	Jan	Feb	March	April	May	June	July	Aug	Sept	Oct	Nov	Dec	Year
Minnesota													
Duluth	47	55	60	58	58	60	68	63	53	47	36	40	55
Minneapolis	49	54	55	57	60	64	72	69	60	54	40	40	56
Iowa													
Des Moines	56	56	56	59	62	66	75	70	64	64	53	48	62
Dubuque	48	52	52	58	60	63	73	67	61	55	44	40	57
Sioux City	55	58	58	59	63	67	75	72	67	65	53	50	63
North Dakota													
Bismarck	52	58	56	57	58	61	73	69	62	59	49	48	59
Devil's Lake	53	60	59	60	59	62	71	67	59	56	44	45	58
Fargo	47	55	56	58	62	63	73	69	60	57	39	46	59
Williston	51	59	60	63	66	66	78	75	65	60	48	48	63
South Dakota													
Huron	55	62	60	62	65	68	76	72	66	61	52	49	63
Rapid City	58	62	63	62	61	66	73	73	69	66	58	54	64
Nebraska													
Lincoln	57	59	60	60	63	69	76	71	67	66	59	55	64
North Platte	63	63	64	62	64	72	78	74	72	70	62	58	68

MINNEAPOLIS, MINNESOTA — Elevation 830 Feet — Table 62

Month	Average Max.	Average Min.	Extreme Max.	Extreme Min.	THI*	Wind Chill Factor*	Precip. Total	Precip. Snow	Not even 0.01" precip.	More than ½" of snow	Clear	Cloudy	Thunder-storms	Fog	90° or higher	32° or lower	% of possible sunshine	Rel. Hum. A.M.	Rel. Hum. P.M.	Wind M.P.H.	Wind Direction	Storm Intensity*
Jan	23	6	58	-31		-16	1	7	23	2	16	15	0	1	0	31	50	81	71	11	NW	S-2
Feb	27	9	56	-28		-13	1	8	21	2	15	13	0	2	0	28	55	81	69	11	NW	S, R-2
March	39	23	78	-27		5	1	11	20	3	15	16	1	1	0	26	53	81	65	12	NW	S, R-2
April	56	36	92	9		21	2	2	21	1	16	14	2	1	0	11	57	76	53	13	NW	R, S-2
May	69	48	95	26			3	T	20	0	16	15	5	1	0	1	57	76	51	12	SE	R-2
June	79	58	100	34	74		4	0	17	0	18	12	8	1	3	0	61	80	56	11	SE	T-3
July	85	63	104	48	77		3	0	21	0	22	9	8	1	7	0	69	82	54	10	SE	T-3
Aug	82	61	102	40	76		3	0	21	0	22	9	6	1	6	0	67	85	56	9	SE	T-3
Sept	73	52	98	26			3	T	21	0	19	11	4	1	1	1	62	85	55	10	SSE	R-2
Oct	60	41	89	18			2	T	23	0	19	12	2	1	0	5	59	80	54	11	SE	R-2
Nov	41	25	75	-9		7	1	7	22	2	13	17	1	1	0	22	39	82	68	12	NW	R, S-2
Dec	27	12	63	-22		-9	1	8	23	2	14	17	0	1	0	30	40	83	73	11	NW	S, R-2
Year	55	36	104	-31			25	43	253	12	205	160	37	13	17	155	56	81	60	11	SE	

Notes:

T Indicates "trace"

* For full explanation of (T-H-I) "Temperature Humidity Index;" "Wind Chill Factor" and "Storm Intensity," see beginning of Chapter 2.

Average date of first freeze .... October 13
" " last " .... April 30
" freeze-free period .... 166 days
10 inches of snow equal approximately one inch of rain.

DULUTH, MINNESOTA Elevation 1409 Feet Table 63

Month	Temperatures Average Max.	Average Min.	Extreme Max.	Extreme Min.	THI*	Wind Chill Factor*	Precip. Total	Precip. Snow	Not even 0.01" precip.	More than ½" of snow	Clear	Cloudy	Thunder-storms	Fog	90° or higher	32° or lower	% of possible sunshine	Rel. Hum. A.M.	Rel. Hum. P.M.	Wind M.P.H.	Wind Direction	Storm Intensity*
Jan	17	−1	52	−35		−32	1	17	20	5	14	17	0	2	0	31	52	77	73	13	NW	S-3
Feb	21	2	46	−29		−26	1	14	18	4	14	14	0	3	0	28	53	79	70	13	NW	S-3
March	31	13	78	−26		−11	2	15	20	4	15	16	1	3	0	30	59	82	68	14	WNW	S, R-3
April	46	28	88	−5		5	3	8	20	2	14	16	1	3	0	22	56	80	60	15	NW	R, S-2
May	60	39	87	20		23	3	1	18	0	16	15	3	5	0	7	57	79	55	14	E	R, S-2
June	70	49	92	30			4	T	17	0	17	13	8	8	0	0	61	85	63	12	E	T-3
July	77	56	97	40	73		4	0	20	0	21	10	8	7	1	0	66	87	62	11	E	T-3
Aug	75	55	97	37	71		3	0	20	0	20	11	8	7	1	0	62	91	63	11	E	T-3
Sept	65	46	89	22			3	T	19	0	14	16	4	4	0	3	50	92	64	12	WNW	R-2
Oct	53	35	86	9		20	2	1	22	0	15	16	2	5	0	12	55	87	61	13	WNW	R, S-2
Nov	34	20	68	−10		−4	2	11	19	3	11	19	0	3	0	26	35	86	76	14	NW	R, S-2
Dec	21	5	50	−33		−21	1	15	20	4	12	19	0	3	0	31	41	80	75	12	NW	S, R-3
Year	47	29	97	−35			30	81	233	22	183	182	35	53	2	190	55	84	66	13	NW	

Notes:

T Indicates "trace"

* For full explanation of (T-H-I) "Temperature Humidity Index;" "Wind Chill Factor" and "Storm Intensity," see beginning of Chapter 2.

Average date of first freeze . . . . October 3
" " " last . . . . May 13
" freeze-free period . . . . 143 days
10 inches of snow equal approximately one inch of rain.

147

DES MOINES, IOWA — Elevation 938 Feet — Table 64

Month	Temperatures Average Max.	Temperatures Average Min.	Temperatures Extreme Max.	Temperatures Extreme Min.	T H I*	Wind Chill Factor*	Precipitation in inches Total	Precipitation in inches Snow	Avg. days Not even 0.01" precip.	Avg. days More than ½" of snow	Avg. days Clear	Avg. days Cloudy	Avg. days Thunder-storms	Avg. days Fog	Avg. days 90° or higher	Avg. days 32° or lower	% of possible sunshine	Rel. Humidity A.M.	Rel. Humidity P.M.	Wind M.P.H.	Wind Direction	Storm Intensity*
Jan	29	11	55	−23		−9	1	9	24	2	14	17	0	2	0	30	51	79	71	12	NW	S, R-2
Feb	32	15	58	−16		−6	1	7	21	2	14	14	0	3	0	27	54	85	74	12	NW	S, R-2
March	43	25	78	−22		7	2	8	21	2	13	18	2	2	0	24	54	86	70	14	NW	R, S-2
April	59	38	88	22		24	3	1	20	1	14	16	4	1	0	8	55	83	60	14	NW	R, S-2
May	71	50	92	32			4	T	20	0	16	15	8	1	0	0	61	82	60	12	SE	T-5
June	81	61	96	45	75		5	0	19	0	18	12	10	1	3	0	66	84	60	11	S	T-5
July	87	65	100	53	79		3	0	22	0	21	10	8	1	8	0	71	84	60	10	S	T-4
Aug	85	63	100	48	78		4	0	22	0	20	11	7	1	5	0	70	87	59	9	S	T-3
Sept	77	54	93	32	71		3	T	22	0	20	10	5	1	1	0	65	88	65	10	S	T-3
Oct	66	43	95	25			2	T	24	0	20	11	3	1	0	4	67	78	52	11	S	R-2
Nov	47	28	76	−3		9	2	2	24	1	16	14	1	2	0	17	54	82	64	12	NW	R, S-2
Dec	34	17	64	−16		−5	1	6	24	2	13	18	0	3	0	29	47	83	71	12	NW	S, R-2
Year	59	39	100	−23			30	33	263	10	199	166	50	18	18	140	60	83	64	11	NW	

Average date of first freeze .... October 19
    "   "   last .... April 20
    "   freeze-free period .... 183 days
10 inches of snow equal approximately one inch of rain.

Notes:
T Indicates "trace"
* For full explanation of (T-H-I) "Temperature Humidity Index;" "Wind Chill Factor" and "Storm Intensity," see beginning of Chapter 2.

FARGO, NORTH DAKOTA Elevation 895 Feet Table 65

Month	Avg Max	Avg Min	Extreme Max	Extreme Min	T H I*	Wind Chill Factor*	Total Precip	Snow	Not even 0.01" precip	More than ½" of snow	Clear	Cloudy	Thunderstorms	Fog	90° or higher	32° or lower	% of possible sunshine	R.H. A.M.	R.H. P.M.	Wind M.P.H.	Wind Direction	Storm Intensity*
Jan	17	−3	53	−36		−39	1	6	22	2	14	17	0	1	0	31	49	73	70	14	NNW	S, R-2
Feb	21	1	43	−30		−32	1	6	21	1	15	13	0	2	0	28	58	77	72	13	NNW	S, R-2
March	35	16	78	−34		−9	1	7	23	2	14	17	0	1	0	29	56	82	71	14	NNW	S, R-2
April	53	31	90	1		10	2	3	22	1	17	13	1	0	0	17	60	82	57	15	NNW	R, S-2
May	67	43	93	17			2	T	21	0	18	13	3	1	0	5	59	78	51	14	NNW	R-2
June	77	53	99	33	72		3	T	19	0	17	13	8	0	2	0	58	84	57	13	SSE	T-3
July	84	58	101	40	77		2	0	21	0	24	7	9	1	5	0	68	86	56	11	SSE	T-3
Aug	82	56	105	37	76		3	0	21	0	22	9	8	1	4	0	64	87	54	12	SSE	T-3
Sept	71	47	99	20			2	T	22	0	18	12	3	1	1	2	56	85	55	13	SSE	R-2
Oct	57	35	90	5		19	1	1	25	0	19	12	1	1	0	12	57	81	56	14	SSE	R-2
Nov	36	19	73	−19		−6	1	5	23	2	12	18	0	1	0	27	39	83	71	14	NNW	R, S-2
Dec	22	4	49	−28		−26	1	7	23	2	14	17	0	2	0	31	46	78	73	13	NNW	S, R-2
Year	52	30	105	−36			19	35	263	10	204	161	33	12	12	182	56	81	62	13	SSE	

Notes:

T Indicates "trace"

* For full explanation of (T-H-I) "Temperature Humidity Index," "Wind Chill Factor" and "Storm Intensity," see beginning of Chapter 2.

Average date of first freeze .... September 26
"   "   last   "   .... May 12
"   freeze-free period   .... 137 days
10 inches of snow equal approximately one inch of rain.

RAPID CITY, SOUTH DAKOTA Elevation 3165 Feet Table 66

Month	Temperatures Average Max.	Average Min.	Extreme Max.	Extreme Min.	T-H-I*	Wind Chill Factor*	Precipitation in inches Total	Snow	Not even 0.01" precip.	More than ½" of snow	Average number of days Clear	Cloudy	Thunder-storms	Fog	90° or higher	32° or lower	% of possible sunshine	Relative Humidity A.M.	P.M.	Wind M.P.H.	Direction	Storm Intensity*
Jan	33	9	74	−27		−10	1	6	25	1	16	15	0	1	0	30	55	71	67	10	NNW	S-2
Feb	36	12	74	−18		−9	T	6	21	3	14	14	0	2	0	27	60	71	64	11	NNW	S-2
March	43	20	82	−15		−2	1	10	22	3	16	15	0	2	0	28	61	74	59	13	NNW	S, R-2
April	57	32	89	9		16	2	5	21	2	15	15	1	2	0	15	62	70	47	13	NNW	R, S-2
May	67	43	93	18			3	1	19	0	17	14	7	1	1	3	56	74	49	12	NNW	T-4
June	76	52	10†	31	70		3	T	17	0	19	11	12	2	3	0	59	75	49	11	NNW	T-5
July	86	59	109	40	77		2	0	22	0	26	5	13	0	12	0	73	70	40	10	NNW	T-3
Aug	85	57	106	40	76		2	0	23	0	26	5	12	1	11	0	70	73	42	10	NNW	T-3
Sept	74	47	103	25	66		1	T	25	0	22	8	4	1	4	1	66	63	39	10	NNW	R-2
Oct	62	36	94	14		23	1	2	26	0	21	10	1	1	0	8	62	64	47	10	NNW	R, S-2
Nov	47	24	77	−14		8	1	4	24	1	16	14	0	1	0	24	57	67	60	11	NNW	S, R-2
Dec	37	14	70	−18		3	T	3	26	1	16	15	0	2	0	30	54	68	65	10	NNW	S-2
Year	58	34	109	−27			17	37	271	11	224	141	50	16	31	166	61	70	52	11	NNW	

Notes:
T Indicates "trace"
* For full explanation of (T-H-I) "Temperature Humidity Index," "Wind Chill Factor" and "Storm Intensity," see beginning of Chapter 2.

Average date of first freeze . . . . October 4
" " " last " . . . . May 7
" freeze-free period . . . . 150 days
10 inches of snow equal approximately one inch of rain.

150

OMAHA, NEBRASKA Elevation 978 Feet Table 67

Month	Average Max.	Average Min.	Extreme Max.	Extreme Min.	THI*	Wind Chill Factor*	Precip. Total	Precip. Snow	Not even 0.01" precip.	More than ½" of snow	Clear	Cloudy	Thunder-storms	Fog	90° or higher	32° or lower	% of possible sunshine	Rel. Hum. A.M.	Rel. Hum. P.M.	Wind M.P.H.	Wind Direction	Storm Intensity*
Jan	32	14	69	−21		−8	1	8	24	3	17	14	0	2	0	30	53	81	68	12	NNW	S, R-2
Feb	37	18	69	−19		−3	1	7	21	2	15	13	0	2	0	27	54	86	67	12	SSE	S, R-2
March	48	28	88	−16		8	1	6	23	2	15	16	2	1	0	21	54	81	60	13	N	R, S-2
April	63	42	91	10			2	1	21	0	16	14	4	1	0	6	58	77	52	14	N	R, S-2
May	73	52	99	29			3	T	20	0	17	14	7	1	2	0	61	80	56	12	SSE	T-4
June	83	62	105	39	77		5	0	19	0	20	10	11	1	8	0	65	81	56	11	SSE	T-5
July	89	68	114	51	81		3	0	22	0	24	7	9	0	15	0	81	81	52	10	SSE	T-3
Aug	86	65	110	43	79		3	0	21	0	23	8	9	1	12	0	70	84	55	10	SSE	T-3
Sept	78	56	104	30	72		3	0	23	0	22	8	5	1	5	0	70	82	51	10	SSE	T-3
Oct	67	44	96	20			2	T	25	0	22	9	3	1	0	4	67	79	50	10	SSE	R-2
Nov	49	29	80	−3		11	1	2	25	1	17	13	1	1	0	19	52	79	58	12	SSE	R, S-2
Dec	36	19	72	−12		1	1	5	26	1	16	15	0	3	0	29	47	81	65	11	SSE	S, R-2
Year	62	41	114	−21			26	31	270	9	224	141	51	15	42	136	62	81	58	11	SSE	

Notes:

T Indicates "trace"

* For full explanation of (T-H-I) "Temperature Humidity Index," "Wind Chill Factor" and "Storm Intensity," see beginning of Chapter 2.

Average date of first freeze .... October 20
    "   "   last .... April 14
    "   freeze-free period .... 188 days
10 inches of snow equal approximately one inch of rain.

mercury up to the 100° mark. T.H.I.'s are in the mid to upper 70s.

Temperatures in Iowa usually reach the upper eighties during the afternoon hours. On occasion, temperatures over 100° are experienced. T.H.I.'s near 80 are the rule, especially during the month of July, and low to mid sixties are commonplace at night.

In the Dakotas almost all the precipitation throughout the summer months falls during thunderstorms. Usually about two such storms can be expected per week. Some are quite severe and it becomes necessary for the weather bureau to issue weather warnings for high winds and damaging hail. Occasionally these storms develop into tornadoes. North Dakota averages two tornadoes a year and may be considered on the fringe of the tornado belt, while South Dakota averages about three.

Minnesota is no different from other midwestern states in that it is characterized by heavy thunderstorms and an occasional tornado throughout the summer months. It averages about two thunderstorms per week during the summer. On the whole, most days are clear and sunny, and overcast days are rare.

June in Iowa and Nebraska, is plagued with considerable poor weather. Violent thunderstorms are quite common (two or three per week). As much as five inches of rain falls during this month, most as a result of these short but severe showers. Nebraska has about a third as many tornadoes as Iowa. The thunderstorms diminish in intensity as the summer progresses through July and August, with rainfall averaging 3″ per month.

Most of this area has a "fair" hay fever rating. The better sections are the northeast corner of Minnesota and the northwest portion of North Dakota.

THE CENTRAL STATES—Area Number 7

See Chart No. 26

(Missouri, Kansas, Arkansas, Oklahoma and Northeast Texas)

Many think of this whole vast area as one big flat grain field, whereas even Kansas, the bread basket of the nation, starts at an elevation of

700 feet in the southeast and climbs to over 4000 feet at its western end in the Rocky Mountains foothills. Although these South Central states are often overlooked by vacationists and retirees, there is one portion which could be particularly attractive to either.

This is the large Ozark Plateau highlands, mostly in Missouri and Arkansas but which extends over into Oklahoma and Kansas. First, however, let's consider the overall area.

GENERAL FEATURES

Very hot, sultry summers; cold waves; violent thunderstorms; winter blizzards, and the home of the tornado—are all characteristics of this overall area. Winters range from cold in Kansas and Missouri to mostly mild in Arkansas, Oklahoma and northeast Texas, where snow is rare. Precipitation amounts are consistent throughout the year in this area except in Kansas where 70% of precipitation occurs during late spring and summer. Rainfalls in Kansas diminish as one heads northwest.

WINTER

Afternoon temperatures in January generally reach the low 50s in Arkansas and mid to upper 40s in Oklahoma. At night the upper 20s and low 30s are the rule. February shows a gradual warming trend. Mid to upper 50s are reached in Arkansas and low 50s in Oklahoma. Cold waves can produce readings near zero and sometimes below. Daytime highs in Kansas and Missouri generally reach the upper 30s to low 40s. Expect temperatures to drop into the teens at night. Cold waves frequently sink the thermometer below zero at night.

Although precipitation is generally in the form of rain, snowfalls are common—especially in the northwest part of Oklahoma. One to two inches per month can be expected in Arkansas and eastern Oklahoma and twice that amount in western Oklahoma. Rainfall is rather heavy during winter. Moderate rains, associated with storms developing in the Gulf of Mexico, wash down the area. Rain or snow can be expected about every third day. Thunderstorms are rare but do occur. Much of the precipitation in eastern Kansas falls as rain, while snow predominates in western Kansas. In December about four inches of snow will

153

fall in eastern Kansas. Almost five inches can be expected during the month of January and between five and six inches in February. Western Kansas receives slightly less.

SPRING AND FALL

Fall readings are three or four degrees warmer than corresponding spring temperatures. In March they will rise to the mid 60s in Arkansas, and near 60° in Oklahoma. Northern Texas experiences figures in the high 60s which drop at night into the chilly 40s in Arkansas and upper 30s in Oklahoma. March temperatures in Kansas and Missouri generally reach into the low to mid 50s and at night drop into the mid 30s. March, however, is a month of drastic changes and averages are misleading. For example, March temperatures in Kansas may occasionally reach the upper eighties and can drop to zero or below within a short period of time. April is more pleasant, and temperature extremes are not as pronounced. May afternoons are in the pleasant 70s and at nighttime in the mid 50s. It is somewhat warmer in the southern sections. Temperatures, however, can reach the 90s and still drop suddenly into the 30s. This weather is marred by the possibility of heavy thunderstorms or an occasional tornado. Autumn is delightful with comfortable temperature and humidity.

During March northern and western Kansas can still expect snow. As indicated above, heavy showers and thunderstorms are common in the spring and begin as early as March in the southern areas and in April for Kansas and Missouri. Such storms occur approximately every third day. Tornadoes are a distinct possibility, especially during late spring; Kansas and Oklahoma are especially susceptible to these violent twisters. They are usually brief although severe and arrive during the afternoon hours. Damaging winds and hail frequently accompany them. Fall precipitation is not as frequent nor as heavy. Thunderstorms decrease in intensity and lighter showers occur about once or twice a week. Weather is generally pleasant during September and October and November begins the snowy season, which is most pronounced in western Kansas.

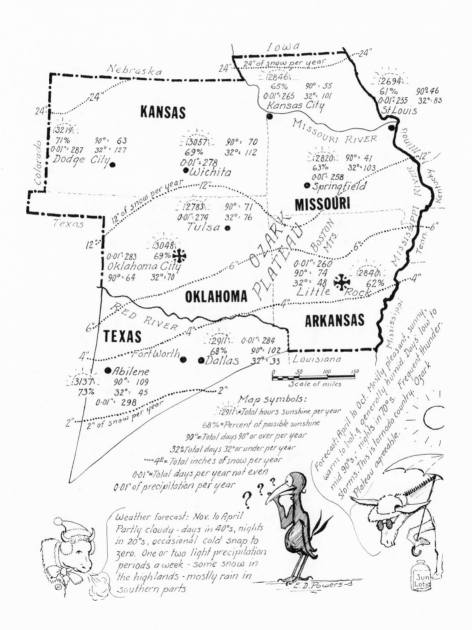

CENTRAL AREA No.7 Chart No. 26

Iowa

Nebraska

KANSAS

24" of snow per year

:2846:
65%   90°: 55
0·01": 265   32°: 101
Kansas City

:2694:
61%   90°: 46
0·01": 255   32°: 83
St.Louis

:3219:
71%   90°: 63
0·01": 287   32°: 127
Dodge City

:3057:
69%   90°: 70
0·01": 278   32°: 112
Wichita

Missouri River

:2820:
63%   90°: 41
0·01": 258   32°: 103
Springfield

Colorado

MISSOURI

12" of snow per year

:2783:
90°: 71
0·01": 274   32°: 76
Tulsa

Texas

:3048:
0·01": 283   69%
Oklahoma City
90°: 64   32°: 70

OZARK PLATEAU

BOSTON MTS.

0·01": 260
90°: 74
32°: 48
Little Rock

:2840:
62%

Mississippi River

Kentucky

Tenn.

6"

4"

OKLAHOMA

RED RIVER

ARKANSAS

TEXAS

:2911:
68%   90°: 102
0·01": 284   32°: 33
Fort Worth   Dallas

Abilene

:3137:
73%   90°: 109
0·01": 298   32°: 45

Louisiana

0   50   100   150
Scale of miles

2"

2" of snow per year

Map symbols:

:2911: = Total hours sunshine per year
68% = Percent of possible sunshine
90° = Total days 90° or over per year
32° = Total days 32° or under per year
4" = Total inches of snow per year
0·01" = Total days per year not even
0·01" of precipitation per year

Forecast: April to Oct. Mostly pleasant, sunny, warm to hot, & generally humid. Days low to mid 90°s, nights in 70°s. Frequent thunder storms. This is tornado country. Ozark Plateau agreeable.

Weather forecast: Nov. to April
Partly cloudy - days in 40°s, nights
in 20°s, occasional cold snap to
zero. One or two light precipitation
periods a week - some snow in
the highlands - mostly rain in
southern parts

? ? ?

E.D. Powers

Jun
Lotio

SUMMER

Summers are very hot and humid. Afternoon temperatures generally reach the nineties, and these temperatures cause much discomfort when accompanied by high humidities. During heat waves, temperatures can soar well over the 100 degree mark. At night they usually remain in the uncomfortable 70s, but do drop into the 60s in Kansas. June and August are slightly cooler than July.

Rainfall is moderate, occurring as thundershowers mainly in afternoon and evening. Many are severe especially in southeastern Kansas and Missouri. Two or three storms per week are probable. As much as five inches of rain falls during the month of June, most of it as a result of these short but violent showers. Tornadoes are very possible during June but the probability of tornadoes decreases as the summer progresses. Thunderstorms are still common. There are usually about 40 or 50 thunderstorms a year in this area.

This is the zone of maximum tornado frequency in the United States. Tornadoes parallel the seasonal trend of thunderstorms and hail. During the 1953–65 period there was an annual average of 77 tornadoes in 28 days in Oklahoma; 72 tornadoes in 27 days in Kansas; 27 in 14 days

Table 68	AVERAGE % OF MAXIMUM POSSIBLE SUNSHINE (% of daylight hours)												
Area Number 7 Central	Jan	Feb	March	April	May	June	July	Aug	Sept	Oct	Nov	Dec	Year
Missouri													
Kansas City	55	57	59	60	64	70	76	73	70	67	59	52	65
Saint Louis	48	49	56	59	64	68	72	68	67	65	54	44	61
Springfield	48	54	57	60	63	69	77	72	71	65	58	48	63
Kansas													
Dodge City	67	66	68	68	68	74	78	78	76	75	70	67	71
Wichita	61	63	64	64	66	73	80	77	73	69	67	59	69
Arkansas													
Little Rock	44	53	57	62	67	72	71	73	71	74	58	47	62
Oklahoma													
Oklahoma City	57	60	63	64	65	74	78	78	74	68	64	57	68
Texas													
Abilene	64	68	73	66	73	86	83	85	73	71	72	66	73
Fort Worth	56	57	65	66	67	75	78	78	74	70	63	58	68

ST. LOUIS, MISSOURI Elevation 535 Feet Table 69

Month	Average Max.	Average Min.	Extreme Max.	Extreme Min.	THI*	Wind Chill Factor*	Precip. Total	Precip. Snow	Not even 0.01″ precip.	More than ½″ of snow	Clear	Cloudy	Thunder-storms	Fog	90° or higher	32° or lower	% of possible sunshine	Rel. Hum. A.M.	Rel. Hum. P.M.	Wind M.P.H.	Wind Direction	Storm Intensity*
Jan	40	24	75	−11		6	2	4	22	2	15	16	1	2	0	27	52	78	60	10	NW	R, S-2
Feb	44	25	83	−2		10	2	4	19	1	15	13	1	1	0	24	51	75	56	11	NW	R, S-2
March	53	32	88	5		20	3	5	20	1	18	13	3	2	0	14	53	77	56	12	WNW	R, S-3
April	66	44	90	23	70		4	T	18	0	17	13	6	0	0	3	54	74	52	11	WNW	T-3
May	75	53	92	31	77		4	T	19	0	20	11	6	0	1	0	63	79	55	10	S	T-4
June	85	63	96	46	81		4	0	19	0	21	9	7	0	7	0	69	82	56	8	S	T-5
July	89	67	106	51	81		3	0	23	0	25	6	7	0	11	0	71	84	57	8	S	T-4
Aug	87	66	105	47	80		3	0	23	0	24	7	6	1	10	0	69	86	55	7	S	T-4
Sept	81	58	96	40	74		3	0	23	0	23	7	3	1	3	0	64	88	56	8	S	T-3
Oct	70	47	94	28			3	T	24	0	23	8	3	0	1	2	65	80	50	9	S	T-2
Nov	54	35	80	1		23	3	1	22	1	18	12	1	1	0	13	50	81	59	10	S	R, S-2
Dec	43	27	73	−6		11	2	3	23	1	15	16	0	1	0	24	45	81	65	10	WNW	R, S-2
Year	66	45	106	−11			35	17	255	6	234	131	43	9	34	106	59	80	56	10	S	R, S-2

Notes:

T Indicates "trace"

* For full explanation of (T-H-I) "Temperature Humidity Index;" "Wind Chill Factor" and "Storm Intensity," see beginning of Chapter 2.

Average date of first freeze . . . . November 8
" " last " . . . . April 2
" freeze-free period . . . . 220 days
10 inches of snow equal approximately one inch of rain.

KANSAS CITY, MISSOURI — Elevation 742 Feet — Table 70

Month	Temperatures Average Max.	Average Min.	Extreme Max.	Extreme Min.	T H I*	Wind Chill Factor*	Precip. Total	Precip. Snow	Not even 0.01" precip.	More than ½" of snow	Clear	Cloudy	Thunder-storms	Fog	900° or higher	32° or lower	% of possible sunshine	Rel. Humidity A.M.	Rel. Humidity P.M.	Wind M.P.H.	Wind Direction	Storm Intensity*
Jan	40	23	73	−9		5	1	6	24	2	16	15	0	2	0	26	53	74	60	10	SSW	R, S-2
Feb	45	27	74	−2		9	1	4	21	1	15	13	1	1	0	22	56	72	57	10	SSW	R, S-2
March	53	34	87	−1		19	2	4	22	1	16	15	3	1	0	14	57	72	54	12	ENE	R, S-2
April	66	46	90	25			4	1	19	0	16	14	5	1	0	2	59	72	51	12	S	T-3
May	75	56	94	34	70		4	T	19	0	18	13	7	1	3	0	63	76	53	11	S	T-4
June	85	67	99	47	77		5	0	19	0	19	11	9	0	6	0	69	82	60	10	S	T-5
July	92	71	104	54	82		3	0	23	0	24	7	8	0	14	0	76	80	58	9	S	T-4
Aug	90	69	104	50	81		4	0	23	0	24	7	7	0	12	0	73	81	57	9	S	T-4
Sept	83	60	97	40	74		3	0	22	0	22	8	5	1	3	0	69	83	58	9	S	T-3
Oct	72	49	97	31			3	T	24	0	21	10	3	1	1	0	68	72	48	9	S	R-2
Nov	55	35	81	7		22	2	1	24	0	18	12	1	1	0	10	58	75	57	10	SSW	R, S-2
Dec	44	28	70	−7		11	2	5	25	1	16	15	0	2	0	24	51	75	62	10	SSW	R, S-2
Year	67	47	104	−9			34	20	265	5	225	140	50	12	39	98	64	76	56	10	S	

Notes:

T Indicates "trace"

* For full explanation of (T-H-I) "Temperature Humidity Index," "Wind Chill Factor" and "Storm Intensity," see beginning of Chapter 2.

Average date of first freeze . . . . October 31
" " last " . . . . April 5
" freeze-free period . . . . 210 days
10 inches of snow equal approximately one inch of rain.

158

LITTLE ROCK, ARKANSAS   Elevation 257 Feet   Table 71

Month	Average Max.	Average Min.	Extreme Max.	Extreme Min.	THI*	Wind Chill Factor*	Precip. Total	Precip. Snow	Not even 0.01" precip.	More than ½" of snow	Clear	Cloudy	Thunder-storms	Fog	90° or higher	32° or lower	% of possible sunshine	Rel. Hum. A.M.	Rel. Hum. P.M.	Wind M.P.H.	Wind Direction	Storm Intensity*
Jan	51	31	78	−4		21	5	3	21	1	14	17	2	3	0	23	45	81	60	9	S	R, S-3
Feb	55	34	83	10		26	4	2	18	0	14	14	2	2	0	17	53	79	56	9	SW	R, S-3
March	63	41	88	17			5	T	21	0	15	16	5	1	0	6	56	77	54	10	WNW	R-3
April	74	51	90	30	76		5	T	19	0	17	13	7	1	0	0	59	79	55	10	S	R-3
May	82	60	98	40	82		5	0	21	0	19	12	7	1	5	0	65	86	57	8	S	T-3
June	90	68	102	49	84		4	0	22	0	21	9	7	0	17	0	71	87	54	8	SSW	T-3
July	93	71	105	56	84		3	0	23	0	22	9	9	1	22	0	69	88	59	7	SW	T-3
Aug	93	70	108	52	83		3	0	24	0	23	8	6	1	18	0	71	88	56	7	SW	T-3
Sept	86	62	102	38	78		3	0	23	0	21	9	4	1	8	0	66	90	59	7	NE	T-3
Oct	76	50	97	31	72		3	0	25	0	22	9	2	2	2	0	70	85	48	7	SW	R-3
Nov	61	38	85	17			4	T	22	0	18	12	3	2	0	6	56	83	57	8	SW	R-3
Dec	52	32	78	−1		23	4	1	22	0	15	16	1	2	0	17	48	80	61	9	SW	R, S-3
Year	73	51	108	−4			49	6	261	1	221	144	56	16	73	68	62	83	56	8	SW	

Notes:

T Indicates "trace"

* For full explanation of (T-H-I) "Temperature Humidity Index," "Wind Chill Factor" and "Storm Intensity," see beginning of Chapter 2.

Average date of first freeze ....November 15
    "   "   " last   " .... March 16
    "   freeze-free period .... 244 days
10 inches of snow equal approximately one inch of rain.

WICHITA, KANSAS   Elevation 1,321 Feet   Table 72

Month	Average Max.	Average Min.	Extreme Max.	Extreme Min.	THI*	Wind Chill Factor*	Precip. Total	Precip. Snow	Not even 0.01" precip.	More than ½" of snow	Clear	Cloudy	Thunder-storms	Fog	90° or higher	32° or lower	% of possible sunshine	R.H. A.M.	R.H. P.M.	Wind M.P.H.	Wind Direction	Storm Intensity*
Jan	41	23	68	−1		3	1	3	25	1	15	16	0	3	0	29	59	78	61	14	S	R, S-2
Feb	48	27	80	4		6	1	3	23	1	14	14	1	4	0	23	63	81	57	14	N	R, S-2
March	56	34	89	6		16	2	1	24	1	16	15	1	1	0	15	66	74	47	15	S	R, S-2
April	67	45	94	24			4	T	23	0	16	14	6	2	1	3	63	78	49	15	S	T-4
May	75	55	99	33	70		4	0	20	0	17	14	10	1	3	0	61	84	56	14	S	T-5
June	86	65	106	44	78		5	0	21	0	20	10	10	0	12	0	75	84	52	14	S	T-5
July	92	69	113	59	82		3	0	23	0	23	8	7	0	27	0	84	74	41	12	S	T-4
Aug	92	68	107	48	82		3	0	24	0	25	6	8	0	25	0	82	72	41	13	S	T-4
Sept	83	61	103	43	76		3	0	23	0	23	7	6	1	12	0	84	71	41	13	S	T-3
Oct	71	50	95	23			2	T	25	0	22	10	4	1	2	2	65	79	53	14	S	T-2
Nov	55	35	80	10		19	2	T	26	0	21	9	0	1	0	15	71	74	49	13	S	R, S-2
Dec	45	26	83	7		6	1	T	26	1	17	14	0	3	0	26	67	77	54	13	NNW	R, S-2
Year	68	46	113	−1			31	8	283	4	228	137	53	17	82	113	70	77	50	14	S	

Notes:

T Indicates "trace"

* For full explanation of (T-H-I) "Temperature Humidity Index;" "Wind Chill Factor" and "Storm Intensity," see beginning of Chapter 2.

Average date of first freeze . . . . .November 1

      "   " last . . . . .April 5

      "   freeze-free period . . . . 210 days

10 inches of snow equal approximately one inch of rain.

OKLAHOMA CITY, OKLAHOMA   Elevation 1,285 Feet   Table 73

Month	Avg Max	Avg Min	Extreme Max	Extreme Min	THI*	Wind Chill Factor*	Precip Total (in.)	Precip Snow (in.)	Not even 0.01" precip	More than ½" of snow	Clear	Cloudy	Thunder-storms	Fog	90° or higher	32° or lower	% of possible sunshine	R.H. A.M.	R.H. P.M.	Wind M.P.H.	Wind Direction	Storm Intensity*
Jan	46	28	79	2		4	1	3	25	1	15	16	0	4	0	22	58	81	64	14	N	R,S-2
Feb	51	31	75	13		11	1	2	21	1	16	12	2	3	0	21	61	80	57	14	N	R,S-2
March	60	38	93	9		20	2	2	24	0	18	13	3	2	1	10	64	76	52	15	SSE	R,S-2
April	71	49	88	28			3	T	22	0	17	13	6	1	0	1	63	79	54	15	SSE	T-2
May	78	59	96	40	75		5	0	20	0	19	12	9	1	3	0	64	82	58	14	SSE	T-3
June	87	69	98	52	81		4	0	21	0	24	6	9	0	13	0	73	84	57	13	SSE	T-3
July	93	72	108	59	84		2	0	24	0	24	7	7	0	24	0	76	81	51	12	SSE	T-3
Aug	94	72	105	54	83		3	0	25	0	26	5	7	0	20	0	78	81	51	11	SSE	T-3
Sept	85	63	93	41	78		3	0	24	0	24	6	5	1	2	0	73	87	57	12	SSE	T-3
Oct	74	52	89	32	69		3	T	25	0	22	9	3	1	0	0	69	78	49	13	SSE	R-3
Nov	59	38	84	17		22	2	T	26	0	22	8	1	2	0	9	62	78	51	13	S	R-2
Dec	49	31	80	1		10	1	2	26	1	19	12	0	3	0	21	59	79	59	13	S	R-2
Year	71	50	108	1			31	10	283	3	246	119	51	18	62	83	67	80	55	13	SSE	

Average date of first freeze . . . . November 6
    "   "   last   " . . . . March 30
    "   freeze-free period . . . . 221 days
10 inches of snow equal approximately one inch of rain.

Notes:
T Indicates "trace"
* For full explanation of (T-H-I) "Temperature Humidity Index," "Wind Chill Factor" and "Storm Intensity," see beginning of Chapter 2.

DALLAS, TEXAS — Elevation 481 Feet — Table 74

Month	Average Max.	Average Min.	Extreme Max.	Extreme Min.	THI*	Wind Chill Factor*	Precip. Total	Precip. Snow	Not even 0.01" precip.	More than ½" of snow	Clear	Cloudy	Thunder-storms	Fog	90° or higher	32° or lower	% of possible sunshine	Rel. Hum. A.M.	Rel. Hum. P.M.	Wind M.P.H.	Wind Direction	Storm Intensity*
Jan	56	36	82	8		24	2	1	24	1	14	17	1	2	0	15	50	77	60	11	S	R-2
Feb	60	39	87	16		27	3	1	20	0	14	14	2	1	0	10	53	76	56	11	S	R-2
March	67	45	91	17			3	T	23	0	18	13	4	1	0	4	58	75	54	13	S	R-2
April	75	55	99	33	72		4	0	21	0	17	13	6	0	1	0	57	79	56	13	S	T-3
May	83	63	98	42	77		5	0	22	0	19	12	7	0	6	0	61	83	58	12	S	T-3
June	91	72	102	53	83		3	0	24	0	24	6	5	0	19	0	73	82	56	12	SSE	T-3
July	95	75	105	65	85		2	0	26	0	24	7	4	0	26	0	77	78	51	10	S	T-2
Aug	95	75	107	61	85		2	0	25	0	26	5	4	0	26	0	77	77	50	10	SSE	T-2
Sept	88	67	103	46	80		3	0	25	0	24	6	3	0	12	0	71	82	56	9	SE	R-3
Oct	79	57	94	38	74		3	0	25	0	22	9	2	1	3	0	67	81	52	9	SSE	R-2
Nov	66	44	86	21			3	T	24	0	20	10	2	1	0	1	62	81	57	10	S	R-2
Dec	58	38	82	10		27	3	T	25	0	17	14	1	2	0	9	55	77	60	10	SSE	R-2
Year	76	56	107	8			35	2	284	1	239	126	41	8	93	39	65	79	55	11	S	

Notes:

T Indicates "trace"

* For full explanation of (T-H-I) "Temperature Humidity Index," "Wind Chill Factor" and "Storm Intensity," see beginning of Chapter 2.

Average date of first freeze ....November 22
   " " last ....March 18
   " freeze-free period ....249 days
10 inches of snow equal approximately one inch of rain.

in Missouri; and 18 in 10 days in Arkansas. There was probably at least one or more a day somewhere throughout Texas during the season.

Those concerned with hay fever should shun this part of the country at pollen time, as ragweed florishes just about everywhere.

While these south central states will never seriously rival California and Florida, the Ozark high country is rapidly becoming a popular vacationland and even now must be rated as one of the major retiree centers of the United States.

It enjoys a very favorable overall climate as shown in the Hot Springs tabulation in this chapter. The famous resort in the Ouachita Mountains experiences the same mild winters as the many other protected sections. It never gets much over one foot of snow a year, which may cover the ground for about four weeks. There are the usual 40 to 50 thunderstorms per year and frequent tornadoes which are quite local in character. There are perhaps five or six days when there is fog and about the same number when hail may be expected. But there is plenty of sunshine—about 55% of the maximum possible in winter and usually 75% during the summer.

All of this may not add up to perfect weather specifications, but the increasing numbers of visitors most certainly attest to its appeal.

Table 68 lists the average percent of maximum possible sunshine in a scattering of spots throughout this area. A comparison will show that it stacks up pretty well against any other in the country. Chart no. 26 shows the total hours of sunshine per year in most of those locations.

THE ROCKY MOUNTAIN STATES
—Area Number 8

See Charts Nos. 27 and 28

(Montana, Idaho, Wyoming, Colorado)

If a committee of sportsmen, outdoor enthusiasts and camera bugs sat down to draw up specifications for an ideal outdoor vacationland, the end result might be a pretty fair description of this magnificent, rugged land.

Much of it is on the top of the world. Montana has an average elevation of only 3400 feet; this because the eastern half of the state is

ROCKY MTS.

E.D.Powers

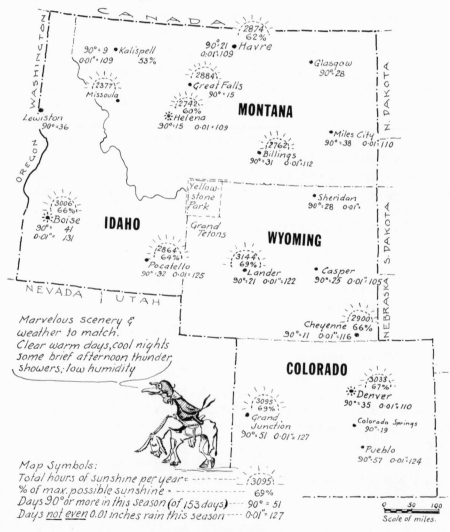

CANADA

WASHINGTON

2874
62%
90°:21 • Havre
0.01"=109

90°=9 • Kalispell
0.01"=109   53%

• Glasgow
90°=28

2377
Missoula

2884
• Great Falls
90°=15

MONTANA

2742
60%
☼ Helena
90°=15 0.01=109

Lewiston
90°=36

OREGON

• Miles City
90°=38 0.01"=110

2762
• Billings
90°=31 0.01"=112

Yellow-
stone
Park

• Sheridan
90°=28 0.01"

N. DAKOTA

S. DAKOTA

3006
66%
☼ Boise
90°= 41
0.01"= 131

IDAHO

Grand
Tetons

2864
64%
• Pocatello
90°=32 0.01"=125

WYOMING

3144
69%
• Lander
90°=21 0.01"=122

• Casper
90°=25 0.01"=105

NEVADA | UTAH

NEBRASKA

Marvelous scenery &
weather to match.
Clear warm days, cool nights
some brief afternoon thunder
showers; low humidity

2900
Cheyenne 66%
90°=11 0.01"=116

COLORADO

3033
67%
☼ Denver
90°=35 0.01"=110

3095
69%
• Grand
Junction
90°=51 0.01"=127

• Colorado Springs
90°=19

• Pueblo
90°=57 0.01"=124

Map Symbols:
Total hours of sunshine per year = - - - - - - -   3095
% of max. possible sunshine = - - - - - - - - -   69%
Days 90° or more in this season (of 153 days) - - -   90° = 51
Days not even 0.01 inches rain this season - - - -   0.01"=127

0   50   100
Scale of miles.

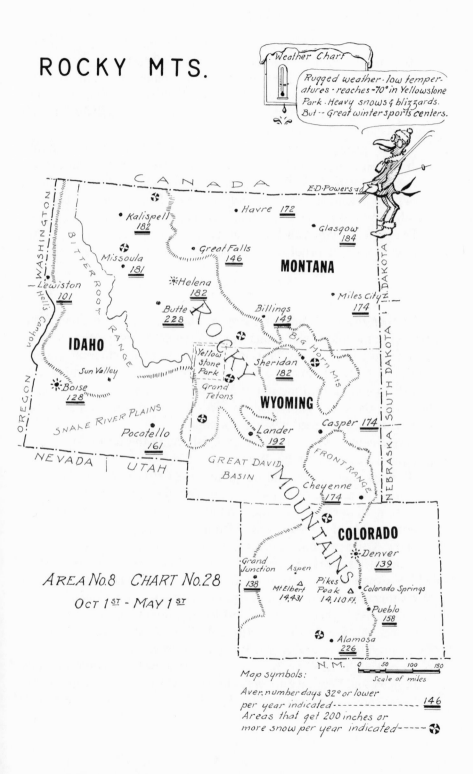

ROCKY MTS.

Weather Chart

Rugged weather · low temperatures · reaches -70° in Yellowstone Park · Heavy snows & blizzards. But · · Great winter sports centers.

CANADA

E·D·Powers

• Havre 172

WASHINGTON

• Kalispell 182

• Glasgow 184

BITTER ROOT RANGE

• Great Falls 146

Missoula 181

MONTANA

☼Helena 182

• Miles City 174

Butte 228

Billings 149

NORTH DAKOTA

Lewiston 101

HELLS CANYON

IDAHO

Yellow Stone Park

Big Horn Mts

Sheridan 182

Sun Valley

Grand Tetons

WYOMING

☼Boise 128

SNAKE RIVER PLAINS

Casper 174

OREGON

Pocatello 161

• Lander 192

FRONT RANGE

NEBRASKA

SOUTH DAKOTA

NEVADA | UTAH

GREAT DAVID BASIN

Cheyenne 174

MOUNTAINS

COLORADO

☼Denver 139

AREA No.8   CHART No.28

OCT 1ST - MAY 1ST

Grand Junction 138

Aspen

△ Mt Elbert 14,431

Pikes Peak △ 14,110 Ft.

Colorado Springs

Pueblo 158

• Alamosa 226

N. M.

0   50   100   150
Scale of miles

Map symbols:

Aver. number days 32° or lower per year indicated- - - - - - - - - - - 146
Areas that get 200 inches or more snow per year indicated- - - - - ❋

low cattle country. Idaho averages 5000 feet and Wyoming is a lofty 6700 feet with 50 mountain peaks over 10,000 feet.

The Continental Divide runs down the center of Colorado whose more than 1000 locations over 10,000 feet high exceed the total number in all the famous Alps. With an average elevation of 6800 feet, it is the highest state in the nation.

All this produces warm days and cool nights during the summer, with frequent but brief afternoon thundershowers blocking off the otherwise sun-filled skies. Winters are long and characterized by heavy snows, high winds and bitterly cold weather. Snow generally begins in October in most places and in many areas doesn't disappear until June.

The popularity of Sun Valley and other skiing and winter activities centers are ample proof that this isn't as bad as it sounds.

WINTER

Temperatures are influenced by altitude to a great degree. Areas located in the higher elevations will experience colder weather than the surrounding lower levels. During January afternoon temperatures generally reach the low to mid 30s in Idaho, Montana and Wyoming, with Colorado getting into the upper 30s and 40s on most days. At night, however, there is usually a drop into the low to mid teens in most places, with lower readings not uncommon.

Severe cold waves strike this area, especially the northern sections. Temperatures in Montana, Wyoming and even Colorado go below the $-30°$ mark quite often, and temperatures of 60 and 70 degrees below zero have been recorded in areas near Yellowstone National Park and in sections of northern Montana. Winds can also be quite intense in this region as frigid air pushes down from Canada.

Snow is the predominant form of precipitation from November on through at least March, except in the southwestern sections of Idaho which experience mostly rain. December shows snow in Montana, Wyoming and most of Colorado, ranging from six to nine inches during the month in higher elevations and lesser amounts in the eastern lowland sections. Heavy snowstorms and blizzards strike the area rather frequently and even when roads are open, extreme caution should be practiced while traveling. In Idaho rain and snow fall in equal amounts.

January and February are very similar to December in the mountains as far as snow is concerned. March is the severest month, with heavier

166

snows and increased probability of blizzards as the warm, moist air from the Gulf of Mexico moves northward to clash with the still bitterly cold weather of the north.

SPRING AND FALL

Fall temperatures are about 5° to 7° higher than in spring.

Expect the mercury to climb into the mid 50s during April in Montana and Wyoming and the low to mid 60s in Idaho and Colorado. At night it will drop into the low to mid 30s in most areas. Cold waves are still possible during April—especially in Montana, Wyoming, and Colorado, with temperatures going down near zero and below in some cases. Idaho is milder with cold waves a rarity and temperatures staying above 20°F. at night, except in northern portions of the state which can be subjected to occasional cold waves. You can expect temperatures in March to be about 9°F or 10° colder than those of April. In May temperatures will be about 9° or 10° warmer than those of April, with cold snaps lowering the thermometer into the teens in some places.

March and April are still snow months in all areas except southwest Idaho. Heavy snow is possible especially during March. Blizzard conditions—blowing snow, cold temperatures and even heavy falling snow—are distinct possibilities. In the higher elevations snow is common in May and many roads are not opened until the month of June. Most places average over one foot of snow per month throughout spring. Expect some precipitation about every third day.

Snow begins falling in September and by October many mountain roads are closed. Fall months, however, do not usually experience the severe blizzards that strike during late winter and spring. Snowfall is common, however, as storms move eastward from the Pacific. Western and southwestern Idaho are spared most of this snow until late November and December; until that time much of the precipitation falls as rain. During the fall months precipitation can usually be expected one or two days a week.

SUMMER

Summers are usually characterized by warm days and cool nights. Afternoon temperatures in the mid to upper 80s are common, with low

167

90s in areas of lower elevations. Low humidities make these temperatures quite comfortable, although the southern lower lands of Idaho can be uncomfortable. Excellent sleeping weather exists here as the temperatures drop into the mid to upper 50s at night in most areas. Occasional heat waves can lift afternoon readings into the high 90s and occasionally over the 100° mark (especially in low elevations) but this is rare.

During the summer one can generally expect pleasantly warm daytime temperatures under sunny skies, followed by clear cool nights.

Table 75 lists the average percent of maximum possible sunshine in almost a dozen places scattered throughout this area. Chart no. 27 shows the total number of hours of sunshine per year at these locations.

Precipitation is in the form of thunderstorms, averaging about thirty a year. These storms, rarely very intense, occur every two or three days, mainly during the afternoon. Areas located in the higher terrain experience more storms. These storms are usually brief and do not deposit a great deal of rainfall as the air carries very little moisture.

Tornadoes are quite infrequent. Through the 1953–65 period Montana had an average of four per year; Idaho two; Wyoming

Table 75	AVERAGE % OF MAXIMUM POSSIBLE SUNSHINE (% of daylight hours)												
Area Number 8 Rocky Mountains	Jan	Feb	March	April	May	June	July	Aug	Sept	Oct	Nov	Dec	Year
Montana													
Havre	49	58	61	63	63	65	78	75	64	57	48	46	62
Helena	46	55	58	60	59	63	77	74	63	57	48	43	60
Kalispell	28	40	49	57	58	60	77	73	61	50	28	20	53
Idaho													
Boise	40	48	59	67	68	75	89	86	81	66	46	37	66
Pocatello	37	47	58	64	66	72	82	81	78	66	48	36	64
Wyoming													
Cheyenne	65	66	64	61	59	68	70	68	69	69	65	63	66
Lander	66	70	71	66	65	74	76	75	72	67	61	62	69
Sheridan	56	61	62	61	61	67	76	74	67	60	53	52	64
Yellowstone	39	51	55	57	56	63	73	71	65	57	45	38	56
Colorado													
Denver	67	67	65	63	61	69	68	68	71	71	67	65	67
Grand Junction	58	62	64	67	71	79	76	72	77	74	67	58	69

BILLINGS, MONTANA — Elevation 3,567 Feet — Table 76

Month	Temperatures Average Max.	Temperatures Average Min.	Temperatures Extreme Max.	Temperatures Extreme Min.	T H I*	Wind Chill Factor*	Precipitation Total	Precipitation Snow	Not even 0.01" precip.	More than ½" of snow	Clear	Cloudy	Thunderstorms	Fog	90° or higher	32° or lower	% of possible sunshine	Rel. Humidity A.M.	Rel. Humidity P.M.	Wind M.P.H.	Wind Direction	Storm Intensity*
Jan	32	13	68	-30		-11	1	7	24	3	15	16	0	2	0	29	49	63	60	13	SW	S-2
Feb	36	16	69	-38		-5	T	9	20	3	13	15	0	2	0	25	53	63	60	12	SW	S-2
March	44	24	75	-19		7	1	11	22	4	13	18	0	2	0	25	56	60	55	12	SW	S-2
April	58	34	92	-5		19	1	7	21	2	14	16	1	2	0	12	59	49	44	12	NE	S, R-2
May	68	44	96	14			2	1	21	1	17	14	5	1	1	2	60	48	43	11	NE	R, S-2
June	77	52	102	32	71		3	T	19	0	19	11	8	1	3	0	65	51	47	11	SW	T-3
July	88	59	106	42	78		1	0	24	0	27	4	8	1	13	0	78	40	32	10	SW	T-2
Aug	86	56	104	40	77		1	0	25	0	26	5	7	1	10	0	75	41	32	10	NE	T-2
Sept	73	47	100	26			1	1	23	0	20	10	2	2	2	1	68	47	39	10	SW	R-1
Oct	61	38	86	4		25	1	3	25	1	20	11	0	2	0	8	63	51	46	11	SW	R, S-2
Nov	45	27	71	-14		9	1	6	24	2	14	16	0	1	0	21	48	59	56	12	SW	S, R-2
Dec	36	18	69	-17		-5	1	8	25	3	15	16	0	2	0	27	47	61	59	13	WSW	S-2
Year	59	36	106	-38			13	53	273	19	213	152	31	19	29	150	62	53	48	11	SW	

Notes:

T Indicates "trace"

* For full explanation of (T-H-I) "Temperature Humidity Index," "Wind Chill Factor" and "Storm Intensity," see beginning of Chapter 2.

Average date of first freeze . . . . September 25
" " last " . . . . May 15
" freeze-free period . . . . 133 days
10 inches of snow equal approximately one inch of rain.

BOISE, IDAHO Elevation 2,842 Feet Table 77

Month	Avg Max	Avg Min	Ext Max	Ext Min	T-H-I*	Wind Chill Factor*	Precip Total	Snow	Not even 0.01" precip.	More than ½" of snow	Clear	Cloudy	Thunderstorms	Fog	90° or higher	32° or lower	% of possible sunshine	R.H. A.M.	R.H. P.M.	Wind M.P.H.	Wind Direction	Storm Intensity*
Jan	35	20	63	−17		6	1	8	19	3	11	20	0	4	0	27	40	75	74	9	SE	S, R-2
Feb	42	26	66	−10		12	1	5	17	2	11	17	0	3	0	23	49	70	66	10	SE	R, S-2
March	52	31	76	10		16	1	2	21	1	14	17	1	1	0	19	58	56	48	11	SE	R, S-2
April	62	38	92	21		25	1	T	22	0	17	13	1	0	0	7	67	48	37	11	SE	R-1
May	71	45	95	27			1	T	22	0	18	13	4	0	1	2	68	47	38	10	NW	T-1
June	78	51	109	34	72		1	T	23	0	22	8	3	0	3	0	75	44	32	10	NW	T-1
July	91	59	106	41	80		T	0	29	0	29	2	3	0	18	0	88	34	23	9	NW	T-1
Aug	88	57	105	41	77		T	0	29	0	28	3	2	0	14	0	86	33	23	9	NW	T-1
Sept	77	48	102	27	70		T	0	27	0	24	6	1	0	4	0	81	38	30	9	SE	T-1
Oct	65	40	88	20		30	1	T	24	0	21	10	1	0	0	5	65	48	43	9	SE	R-1
Nov	49	30	73	−3		18	1	2	20	1	13	17	0	3	0	19	44	67	66	9	SE	R, S-2
Dec	31	24	62	−1		12	1	5	20	2	11	20	0	6	0	25	35	76	77	9	SE	R, S-2
Year	62	39	109	−17			11	22	275	9	219	146	16	17	40	127	66	53	46	10	SE	

Average date of first freeze .... October 12
  " " " last .... May 5
  "   freeze-free period ....160 days
10 inches of snow equal approximately one inch of rain.

Notes:

T Indicates "trace"

* For full explanation of (T-H-I) "Temperature Humidity Index," "Wind Chill Factor" and "Storm Intensity," see beginning of Chapter 2.

CHEYENNE, WYOMING — Elevation 6,131 Feet — Table 78

Month	Average Max.	Average Min.	Extreme Max.	Extreme Min.	THI*	Wind Chill Factor*	Precip. Total	Precip. Snow	Not even 0.01" precip.	More than ½" of snow	Clear	Cloudy	Thunderstorms	Fog	90° or higher	32° or lower	% of possible sunshine	R.H. A.M.	R.H. P.M.	Wind M.P.H.	Wind Direction	Storm Intensity*
Jan	37	14	63	−27		−12	1	7	25	2	19	12	0	1	0	30	60	48	55	14	W	S-2
Feb	40	16	70	−34		−11	1	6	21	2	16	12	0	2	0	27	64	48	54	15	W	S-2
March	44	20	73	−21		−5	1	12	21	4	18	13	0	2	0	28	62	48	51	15	WNW	S-2
April	54	29	78	−6		9	2	11	20	3	16	14	3	3	0	19	57	47	48	14	WNW	S-2
May	63	37	89	16		23	2	5	18	1	16	15	8	3	0	6	57	47	49	12	SSE	R, S-2
June	74	47	100	25	67		2	1	19	0	21	9	12	1	1	1	65	42	44	11	W	T-3
July	83	54	100	38	73		2	0	20	0	25	6	14	1	5	0	69	36	40	10	W	T-3
Aug	81	53	96	37	73		2	0	21	0	23	8	11	1	3	0	66	36	40	10	W	T-3
Sept	72	43	92	18			1	1	23	0	22	8	5	2	0	2	71	37	39	10	W	R-2
Oct	60	33	82	2		19	1	3	25	1	23	8	1	3	0	12	69	39	45	11	W	R, S-2
Nov	47	23	73	−12		0	1	8	24	3	20	10	0	2	0	26	63	45	54	14	W	S-2
Dec	40	17	69	−16		−9	1	5	26	2	19	12	0	2	0	28	59	47	56	14	W	S-2
Year	58	32	100	−34			16	58	263	18	238	127	54	23	9	179	64	43	48	12	W	

Average date of first freeze .... September 15
" " " last " .... May 15
" freeze-free period .... 130 days
10 inches of snow equal approximately one inch of rain.

Notes:

T Indicates "trace"

* For full explanation of (T-H-I) "Temperature Humidity Index;" "Wind Chill Factor" and "Storm Intensity," see beginning of Chapter 2.

DENVER, COLORADO   Elevation 5,292 Feet   Table 79

Month	Temperatures Average Max.	Average Min.	Extreme Max.	Extreme Min.	THI*	Wind Chill Factor*	Precip. Total	Precip. Snow	Not even 0.01" precip.	More than ½" of snow	Clear	Cloudy	Thunderstorms	Fog	90° or higher	32° or lower	% of possible sunshine	RH A.M.	RH P.M.	Wind M.P.H.	Wind Direction	Storm Intensity*
Jan	42	15	65	-25		-1	1	9	25	3	20	11	0	1	0	30	73	43	47	10	S	S-2
Feb	45	18	76	-18		3	1	8	22	2	17	11	0	2	0	26	71	45	46	10	S	S-2
March	50	23	83	-4		9	1	13	23	4	19	12	0	1	0	26	71	42	41	10	S	S, R-2
April	61	32	84	13		20	2	9	21	3	16	14	1	1	0	14	65	35	31	11	S	S, R-2
May	71	42	91	26			3	2	20	0	18	13	6	1	1	2	64	37	36	10	S	R, S-2
June	82	51	96	40	73		1	T	21	0	22	8	10	1	4	0	70	42	40	9	S	T-2
July	88	57	99	48	76		2	0	22	0	25	6	11	0	17	0	72	36	36	9	S	T-2
Aug	87	56	100	41	75		1	0	23	0	24	7	8	1	9	0	72	37	36	9	S	T-2
Sept	79	47	97	29	68		1	2	24	0	23	7	3	1	2	1	75	40	36	8	S	R-2
Oct	67	36	87	19		27	1	3	25	1	23	8	1	1	0	7	76	30	30	8	S	R, S-2
Nov	52	24	74	-2		12	1	7	25	2	21	9	0	1	0	25	66	42	48	9	S	S, R-2
Dec	45	18	71	-16		1	1	6	27	2	22	9	0	1	0	30	68	44	51	9	S	S-2
Year	64	35	100	-25			15	58	279	17	250	115	41	11	32	162	70	39	40	9	S	

Notes:

T Indicates "trace"

* For full explanation of (T-H-I) "Temperature Humidity Index," "Wind Chill Factor" and "Storm Intensity," see beginning of Chapter 2.

Average date of first freeze .... October 14
    "   " " last .... May 2
    " freeze-free period .... 165 days
10 inches of snow equal approximately one inch of rain.

six, and Colorado was tops with 15.

Travelers concerned with hay fever will be interested to know that this whole area rates very good to excellent, with few exceptions. There is, however, considerable air pollution around Denver and some of the mining areas. Lander, Wyoming is also not recommended. It should be remembered that sagebrush pollen may be quite as annoying as ragweed to many with allergies.

NEVADA AND UTAH—Area Number 9

See Charts Nos. 29 and 30

Elevation has a marked influence on climate and these two states have some of the highest rooftops in the country. Nevada, with an average elevation of 5500 feet, is the fifth highest state and Utah's 6100 feet makes it the third. The average of the whole United States is about 2500 feet.

For all their height, Nevada and much of western Utah are in an immense depression called the Great Basin. The towering Sierra Nevadas wall it in on the west and the Wasatch and first ranges of the Rockies to the east, forming a 7000 to 10,000 foot high fencing.

As might be expected from their latitude and elevation, these states generally have hot summers, cold winters and wide daily temperature ranges. There is an overall lack of rainfall and Nevada, with less than nine inches per year, is the driest state in the Union. This is especially true of the western sections which lie in the lee of the moisture draining Sierra Nevada Mountains. As might also be assumed, there is plenty of sunshine. This is illustrated in table 80 covering average percent of maximum possible sunshine—and also on chart no. 30, which shows the total hours of sunshine per year at these same locations. It should be noted that this area is rather free from strong winds. The expectancy of tornadoes is slight. Between 1953 and 1965, Utah experienced an average of only one a year and Nevada had less than one annually.

WINTER

As a general rule during January temperatures range as follows:

Northern areas— Maximum 35–40° dropping to near 10° at night.

NEVADA
UTAH

AREA No.9   CHART No.29
MAY 1ˢᵀ to NOV. 1ˢᵀ

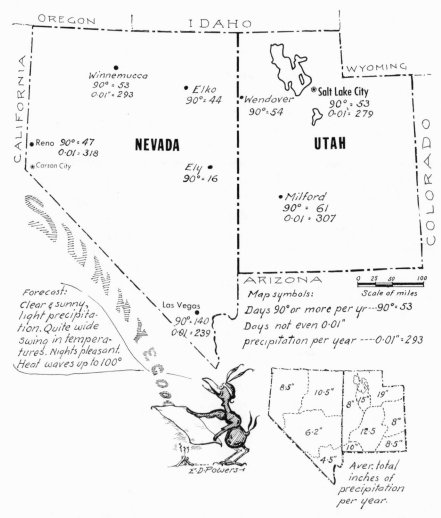

OREGON | IDAHO

CALIFORNIA

WYOMING

Winnemucca
90° = 53
0·01" = 293

• Elko
90° = 44

•Wendover
90°=54

⊛ Salt Lake City
90° = 53
0·01 = 279

• Reno 90° = 47
0·01 = 318

⊛ Carson City

NEVADA

UTAH

COLORADO

Ely •
90° = 16

• Milford
90° = 61
0·01 = 307

ARIZONA

Scale of miles
0   25   50   100

Forecast:
Clear & sunny,
light precipita-
tion. Quite wide
swing in tempera-
tures. Nights pleasant.
Heat waves up to 100°

Las Vegas
90°= 140
0·01 = 239

Map symbols:
Days 90° or more per yr ---90° = 53
Days not even 0·01"
precipitation per year ----0·01" = 293

8·5"   10·5"
8" 15"   19"

6·2"
12·5   8"

10"
8·5"

4·5"

Aver. total
inches of
precipitation
per year.

I.D.Powers

NEVADA
UTAH

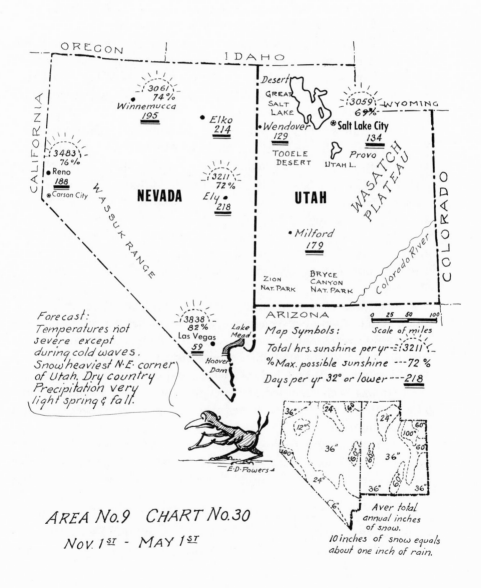

OREGON   IDAHO

CALIFORNIA

3061
74%
Winnemucca
195

Elko
214

Desert
GREAT
SALT
LAKE

Wendover
129

3059
69%
WYOMING

Salt Lake City

134

TOOELE
DESERT

Provo
UTAH L.

3483
76%
Reno
188

Carson City

NEVADA

3211
72%
Ely
218

UTAH

WASATCH PLATEAU

COLORADO

WASSUK RANGE

Milford
179

ZION
NAT. PARK

BRYCE
CANYON
NAT. PARK

Colorado River

Forecast:
Temperatures not
severe except
during cold waves.
Snow heaviest N·E· corner
of Utah. Dry country
Precipitation very
light spring & fall.

3838
82%
Las Vegas
59

Lake
Mead

Hoover
Dam

ARIZONA

Map Symbols:

Scale of miles
0 25 50 100

Total hrs. sunshine per yr --- 3211
% Max. possible sunshine --- 72%
Days per yr 32° or lower --- 218

E·D·Powers

36" 24" 60"
12"   24" 60"
                100"
         36"   60"
100"   60" 60" 36"
  24"   60"
        36"   36"
  .6"

Aver. total
annual inches
of snow.
10 inches of snow equals
about one inch of rain.

AREA No.9   CHART No.30
Nov. 1ST - MAY 1ST

Central areas— Afternoons upper 30s to low 40s and at night drop into the low teens

Southern areas— Afternoon temperatures reach mid to upper 50s and at night drop into the teens in some areas and 20s to 30s in others.

Temperature variations are greatly affected by altitude, as can be seen by referring to the Climatic Tables of Las Vegas at 2161 feet, and Reno and Salt Lake City some 2000 feet higher.

December and February are a few degrees warmer than January, but it must be remembered that cold waves are capable of plunging the thermometer well below zero in the northern parts of both Utah and Nevada. Even in the southern sections, such as Las Vegas, temperatures in the teens have been recorded.

Virtually all winter precipitation falls as snow in Utah and Nevada, except for the extreme southern sections. Amounts vary greatly, with topography the controlling factor. Places in the western part of Nevada especially receive very little precipitation because they are in the shadow formed by the Sierra Nevadas. Generally only one to two inches of total precipitation, (measured as rain) is received during the winter months. Since this is usually almost all snow, about five to eight inches of snow can be expected each month, with occasional storms dropping much more. (Note: one inch of rain equals about ten inches of snow.)

As one travels eastward through Nevada and then into Utah, precipitation amounts vary between one inch to two inches depending on local topography, with higher regions receiving more than those at lower elevations. Heaviest snowfall is recorded in the extreme northeast corner of Utah in the Wasatch Range where as much as 346 inches of snow per year has been recorded at the higher elevations. Even Salt Lake City, located on the western slopes of this range receives over 50 inches of snow annually.

The season actually begins in September in the northern sections and by December snowfalls are common in almost all of Nevada and Utah except extreme southern sections. Even Las Vegas experiences snow in January.

Many roads in Bryce National Park are closed because of snow. In Zion National Park, with its lower elevations and more southern location, roads are kept open all year. Although snow may fall intermittently from December to March, it usually lasts only a day or two on the canyon floor.

In April and October afternoon temperatures generally reach the low to mid 60s and drop into the mid 30s at night. Afternoon temperatures in March and November are in the 50s and drop into upper twenties at night. In September and May afternoon temperatures can be expected in the 70s dropping to the 40s at night. Temperatures can be quite extreme during spring and fall, reaching as high as the 80s and 90s and dropping into the teens during cold snaps in the northern sections. In extreme southern parts such as Las Vegas, the mercury can climb to over 100° but can also drop into the 30s.

Precipitation is generally light totaling between two and four inches in most places. March and November, however, are still snow months; six to eight inches of snow during each is quite common in many areas, especially Utah. Nevada is spared some of this precipitation. During April and October northern areas and those at high elevations still experience snowfalls; however, these additional accumulations are usually quite small. Throughout May and September very little precipitation occurs. Rain can be expected on an average of once a week or less but it is generally very light. Traces of snow are possible only in the high elevations. Thundershowers become a distinct probability during afternoon hours of these months.

Afternoon temperatures in the northern areas climb into the low 90s but drop into the comfortable 50s at night (good sleeping weather). Temperatures can go over 100° during heat waves, while in cold snaps, the morning thermometer can drop into the low 40s, and even some 30s have been registered.

In the central areas afternoon temperatures range into the upper 80s and low 90s. This heat combined with the low humidities creates very comfortable weather. However, occasional heat waves can drive the temperature over the 100° mark, where most people will feel a bit uncomfortable.

In southern Nevada afternoon temperatures rise into the high 90s to over the 100° mark. Even though the air is dry, the extreme heat in the desert areas such as Las Vegas and Beatty cause considerable discomfort, and air conditioning is suggested both in living quarters and cars. Temperatures as high as 116° have been recorded in some of these areas. But then it must be remembered that this is only a short distance from that famous hot spot—Death Valley, which extends into Nevada. That is a place to avoid in summer! For no good reason we once found ourselves covering mile after lonely mile of its scorching bare wastes. After hours of weary driving, we spotted a single little patch on the horizon that aroused our curiosity. It turned out to be what appeared to be a huge blank billboard. Approaching closer we could read three words in small print on the otherwise bare surface. They were:

MONOTONOUS—AINT IT?

Precipitation is light during the summer months throughout all of Nevada and Utah and is generally in the form of afternoon thundershowers, perhaps one a week. These are usually in or around higher terrain. Sunshine is usually abundant during the summer months, visibility excellent, and conditions for vacationing are almost perfect.

To sum up we might say that winters are agreeably mild in the desert and lower parts and cold in the mountains and higher elevations. Summers are pleasantly cool in the high country and hot in the lower parts. Not a bad parlay.

Nevada consists largely of highlands with lofty mountains on the west and partly to the east. The northwest is desert and the southern almost semi-tropical where large crops of cotton, grapes, pomegranates,

Table 80	AVERAGE % OF MAXIMUM POSSIBLE SUNSHINE (% of daylight hours)												
Area Number 9 Nevada-Utah	Jan	Feb	March	April	May	June	July	Aug	Sept	Oct	Nov	Dec	Year
Nevada													
Ely	61	64	68	65	67	79	79	81	81	73	67	62	72
Las Vegas	74	77	78	81	85	91	84	86	92	84	83	75	82
Reno	59	64	69	75	77	82	90	89	86	76	68	56	76
Winnemucca	52	60	64	70	76	83	90	90	86	75	62	53	74
Utah													
Salt Lake City	48	53	61	68	73	78	82	82	84	73	56	49	69

RENO, NEVADA Elevation 4,404 Feet Table 81

Month	Average Max.	Average Min.	Extreme Max.	Extreme Min.	THI*	Wind Chill Factor*	Precipitation Total	Precipitation Snow	Not even 0.01" precip.	More than ½" of snow	Clear	Cloudy	Thunderstorms	Fog	90° or higher	32° or lower	% of possible sunshine	RH A.M.	RH P.M.	Wind M.P.H.	Wind Direction	Storm Intensity*
Jan	46	17	68	−16		11	1	7	25	3	17	14	0	2	0	28	65	67	54	7	S	S, R-2
Feb	51	22	70	−9		16	1	4	22	2	16	12	0	1	0	26	68	60	51	6	S	S, R-2
March	57	25	77	−2		15	1	5	25	2	18	13	0	0	0	28	68	45	35	8	WNW	S, R-2
April	66	30	87	13		21	T	1	26	0	20	10	1	0	0	19	74	37	30	8	WNW	R-1
May	73	37	94	20		29	T	T	26	0	22	9	2	0	1	7	75	34	29	8	WNW	R-1
June	81	42	100	25	71		T	T	27	0	26	4	2	0	4	2	81	31	24	8	WNW	R-1
July	92	47	104	35	77		T	0	28	0	29	2	4	0	20	0	91	25	19	7	WNW	T-1
Aug	90	45	102	32	76		T	0	29	0	29	2	2	0	15	0	92	26	18	7	WNW	T-1
Sept	82	39	101	21	71	35	T	T	28	0	27	3	1	0	7	3	90	31	21	6	WNW	T-1
Oct	70	31	88	13		28	1	T	28	0	23	8	1	0	0	19	77	43	31	6	WNW	R-1
Nov	57	23	77	7		19	1	2	26	1	19	11	0	1	0	27	69	54	43	5	S	R, S-2
Dec	47	19	69	−5		15	1	4	25	1	16	15	0	4	0	29	58	67	58	5	S	S, R-2
Year	68	31	104	−16			7	23	318	9	262	103	13	8	47	188	77	43	34	7	WNW	

Notes:

T Indicates "trace"

* For full explanation of (T-H-I) "Temperature Humidity Index;" "Wind Chill Factor" and "Storm Intensity," see beginning of Chapter 2.

Average date of first freeze . . . . October 2
    "   "   last   " . . . . May 14
    "   freeze-free period . . . . 141 days
10 inches of snow equal approximately one inch of rain.

LAS VEGAS, NEVADA   Elevation 2,162 Feet   Table 82

Month	Temperatures Average Max	Average Min	Extreme Max	Extreme Min	THI*	Wind Chill Factor*	Precip. Total	Snow	Not even 0.01" precip.	More than ½" of snow	Clear	Cloudy	Thunder-storms	Fog	90° or higher	32° or lower	% of possible sunshine	Rel. Hum. A.M.	P.M.	Wind M.P.H.	Direction	Storm Intensity*
Jan	55	33	76	8		26	T	1	28	1	19	12	0	0	0	19	74	41	34	7	W	R, S-1
Feb	62	39	82	17		31	1	T	30	0	20	8	0	0	0	11	77	33	24	8	W	R-1
March	69	44	89	19			T	T	29	0	23	8	0	0	0	3	79	25	19	8	SW	R-1
April	79	53	99	31			T	0	28	0	24	6	1	0	4	0	80	21	16	10	SW	R-1
May	88	60	109	38	70		T	0	30	0	27	4	1	0	17	0	86	17	12	11	SW	T-1
June	99	68	116	48	79		T	0	29	0	28	2	1	0	26	0	92	13	10	11	SW	T-1
July	105	76	117	56	84		T	0	28	0	28	3	5	0	31	0	85	19	14	10	SW	T-1
Aug	103	74	116	54	83		T	0	29	0	29	2	3	0	30	0	86	20	15	9	SW	T-1
Sept	96	65	113	43	78	32	T	0	29	0	28	2	1	0	25	0	92	17	12	8	SW	R-1
Oct	82	53	100	32			T	T	29	0	28	3	1	0	7	0	84	24	19	7	SW	R-1
Nov	67	41	84	15			T	T	28	0	25	5	0	0	0	8	82	29	25	6	W	R-1
Dec	58	36	78	14			1	T	29	0	21	10	0	0	0	17	73	37	31	6	W	R-1
Year	80	53	117	8			4	1	342	1	300	65	13	0	140	58	83	25	19	9	SW	

Notes:

T Indicates "trace"

* For full explanation of (T-H-I) "Temperature Humidity Index," "Wind Chill Factor" and "Storm Intensity," see beginning of Chapter 2.

Average date of first freeze ....November 13
  " " " last " ....March 13
  " " freeze-free period ....245 days
10 inches of snow equal approximately one inch of rain.

180

SALT LAKE CITY, UTAH   Elevation 4,220 Feet   Table 83

Month	Average Max.	Average Min.	Extreme Max.	Extreme Min.	THI*	Wind Chill Factor*	Precip. Total	Precip. Snow	Not even 0.01" precip.	More than ½" of snow	Clear	Cloudy	Thunder-storms	Fog	900° or higher	32° or lower	% of possible sunshine	Rel. Hum. A.M.	Rel. Hum. P.M.	Wind M.P.H.	Wind Direction	Storm Intensity*
Jan	36	17	60	-22		5	1	14	21	4	14	17	0	4	0	28	46	70	72	7	SE	S-2
Feb	43	24	68	-30		14	1	9	19	3	13	15	1	2	0	23	53	64	63	8	SE	S, R-2
March	52	30	78	5		18	2	8	21	3	17	14	1	0	0	20	61	51	49	9	SE	S, R-2
April	63	37	85	14		27	2	3	21	1	17	13	2	0	0	7	67	42	39	9	SE	R, S-2
May	73	45	93	27			2	T	23	0	21	10	5	0	1	1	72	37	32	9	SE	T-2
June	82	52	103	35	73		1	T	25	0	24	6	5	0	7	0	78	31	27	9	SE	T-2
July	92	61	106	41	79		1	0	27	0	27	4	7	0	22	0	82	27	23	10	SE	T-2
Aug	90	59	103	39	77		1	0	25	0	27	4	8	0	18	0	81	28	24	10	SE	T-2
Sept	79	49	98	30	70		1	T	25	0	26	4	4	0	4	0	84	30	27	9	SE	T-2
Oct	67	39	88	18		29	1	1	25	0	23	8	2	0	0	5	72	42	43	9	SE	R-2
Nov	50	29	74	-14		19	1	6	23	2	18	12	0	1	0	22	55	58	65	8	SE	R, S-2
Dec	40	23	66	-21		15	1	11	22	3	15	16	0	3	0	27	43	69	74	8	SE	S, R-2
Year	64	39	106	-30			15	52	279	17	242	123	35	10	52	133	68	46	45	9	SE	

Notes:

T Indicates "trace"

* For full explanation of (T-H-I) "Temperature Humidity Index," "Wind Chill Factor" and "Storm Intensity," see beginning of Chapter 2.

Average date of first freeze . . . . November 1

     "   "   "   last . . . . April 12

     "   freeze-free period . . . . 202 days

10 inches of snow equal approximately one inch of rain.

181

figs and almonds are produced with much irrigation.

Utah's north and east are plateaus indented with valleys. The central area is mountainous and the west desert.

Those concerned with hay fever will note that conditions in both states are generally good to excellent, except at the Canyon rim section of Utah.

THE SOUTHWEST—Area Number 10

See Chart No. 31

(Arizona, New Mexico, Western Texas)

Gold is where you find it and here indeed is a glittering golden land, many parts of which are drenched in brilliant yellow sunshine more than 90% of the maximum possible daylight hours. This is the lower end of the western high country, Arizona having an average elevation of 4100 feet and New Mexico, 5700 feet. A soft, clear atmosphere and not too much rainfall attracts, an ever increasing number of visitors to Arizona in particular and makes it one of the fastest growing states in the nation.

Late spring, summer and early fall in the mountainous areas and late fall, winter and early spring in the desert country are equally perfect. Quite a combination!

Summers are pleasantly mild or sizzlingly hot, and the winters agreeably cool or sharply cold, depending upon the altitude. Daily temperature ranges can be wide. As a rule more rain falls in summer than winter but there are exceptions. There is heavy snow in the mountains but almost never in the desert lowlands. Most of the 30 to 40 thunderstorms per year occur from July through September. The northesat corner of New Mexico may get as many as 70 thunderstorms annually.

Tornadoes are quite uncommon and during the 1953–62 period Arizona averaged about three a year and New Mexico nine. One can occur somewhere in huge Texas almost daily throughout the summer.

Arizona and New Mexico enjoy good to excellent hay fever ratings generally. But Texas is infested with ragweed.

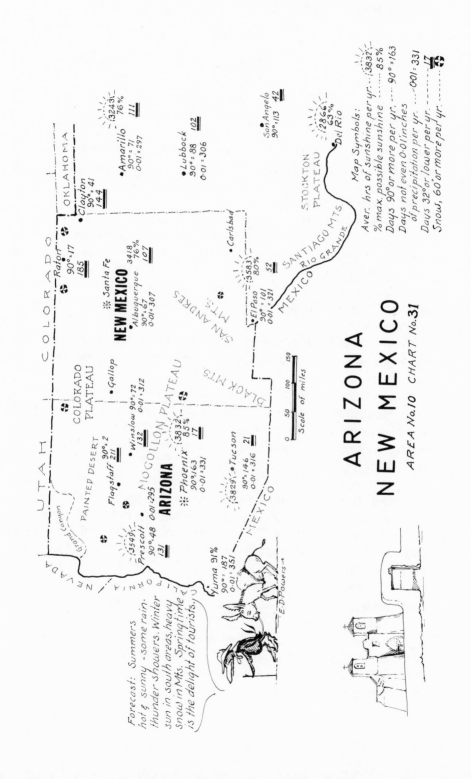

ARIZONA
NEW MEXICO

AREA No.10 CHART No.31

UTAH

COLORADO

OKLAHOMA

NEVADA

ARIZONA

NEW MEXICO

MEXICO

Grand Canyon

PAINTED DESERT

COLORADO PLATEAU

MOGOLLON PLATEAU

BLACK MTS

SAN ANDRES MTS.

SANTIAGO MTS.

STOCKTON PLATEAU

MEXICO

Rio Grande

Scale of miles
0   50   100   150

Yuma 91%
90°·187
0·01·35!

·3549·
Prescott
90°·48
131

Flagstaff 90°·2
211

·Gallop

Winslow 90°·72
132   0·01·312

·3832·
Phoenix 85%
90°·63   17
0·01·331

·3829·Tucson
90°·146   21
0·01·316

Raton
90°·17
185

❀ Santa Fe

·3243·
76%
Amarillo   111
90°·71
0·01·297

Clayton
90°·41
144

·Lubbock
90°·88   102
0·01·306

Albuquerque 3418
90°·67   76%
0·01·307   107

·Carlsbad

·3583·
El Paso   80%
90°·101   52
0·01·321

San Angelo
90°·113   42

·2866·
63%
Del Rio

Map Symbols:

Aver. hrs of sunshine per yr. ·3832·
% max. possible sunshine 85%
Days 90° or more per yr. 90°·163
Days not even 0·01 inches
of precipitation per yr. 0·01·331
Days 32° or lower per yr. 17
Snow, 60 or more per yr. ❀

Forecast: Summers
hot & sunny - some rain-
thunder showers. Winter
sun in south areas, heavy
snow in Mts. Springtime
is the delight of tourists.

E.D.Powers

Winter

Temperature depends largely upon elevation. In the northern half of this area (Colorado Plateau) afternoon temperatures during January reach the low 40s and drop into the teens at night. High elevations experience readings in the low teens. In the Grand Canyon area, afternoon temperatures reach the 40s and sometimes the 50s but drop rapidly below freezing at night. December and February experience slightly warmer temperatures and can drop below zero at night during cold waves—especially in the higher terrain where readings as low as $-30°$ have been recorded.

In the lower elevations, south of the Colorado plateau, temperatures are somewhat milder. During January, afternoon temperatures can be expected to reach the low to mid 60s. At night, however, because of the dryness, temperatures drop rapidly into the 30s. December and February are a few degrees warmer. Portions of southeast New Mexico and western Texas experience daytime temperatures in the mid 50s during this period.

Temperature fluctuations are great in this area, with occasional winter warm spells bringing the temperature into the eighties, and cold spells dropping the mercury to near 20°F.

Precipitation in the plateau area is generally light, but starting in early November it is mostly in the form of snow. Higher elevations in the plateau usually accumulate about 40 to 60" of snow during the winter. Several heavy snowstorms can be expected in the high elevations. Lower elevations receive only about 20". For example, the Grand Canyon area gets about 10" to 20" of snow, but generally the ground is covered from 40–80 days during the winter months. Light snow or rain can be expected about one day a week in the Plateau area.

Precipitation is light in Tucson, Phoenix, El Paso and areas south and east of the plateau and falls in the form of rain. Traces of snow can be found, but it melts almost immediately and the ground may be snow-covered perhaps one day a year. Here again, precipitation may be expected once a week on the average. Generally the skies are sunny and the weather is pleasant.

Western Texas also experiences little snowfall except for the panhandle area. Here developing storms drop between 10 to 20" of snow per year and the ground is usually snow-covered anywhere from 10 to 20 days during the winter.

Spring and Fall

Fall temperatures in the plateau area are several degrees warmer than in spring. During April afternoon temperatures generally reach the 60° mark and drop to near 30° at night. Areas of lower elevations reach temperatures in the upper 60s and low 70s and drop into the 40s at night. March is about 8–10° cooler than April and May is about 7–10° warmer than April. Spring is sometimes punctuated by cold snaps.

South and east of the plateau fall readings are several degrees warmer than corresponding spring temperatures. By April, temperatures in most southern sections can be expected to reach into the pleasant low to mid 80s. Low humidities create ideal conditions for comfort under brilliantly sunny skies. At night ideal sleeping conditions exist as temperatures drop back rapidly to near 50°.

March is 6–8° cooler than April, and May is 6 to 8° warmer than April. In May temperatures can soar to the 90 degree mark and during warm spells over 100° have been recorded. Even though the air is dry, the heat can be uncomfortable.

Moving eastward into southeast New Mexico and western Texas, temperatures are usually in the mid to upper seventies during the day in April, dropping to near 50° at night. The panhandle area of Texas is somewhat cooler and is threatened by cold waves during spring which occasionally drop temperatures into the teens. Heat waves, however, can lift the mercury into the 90s.

Generally 2″ to 4″ of precipitation can be expected in the plateau region in spring and fall. During March and early April snow is not unusual, especially in the higher elevations. Through April and May the snow in the hills melts and the natural rock formations become more visible. As spring progresses, the slight possibility of thunderstorms increases.

In the fall, snow begins in October and November especially in the high areas. Generally by November light accumulations are probable. Spring precipitation south and east of the Plateau is especially dry in this area. A few light rain showers may occur once every two weeks, but nothing more serious. During the fall slightly more rain may fall because of a few afternoon thundershowers, especially during September and early October. But again, problems created by rainy weather are non-existent, and sunshine is abundant.

Table 84 lists the average percentage of maximum possible sunshine throughout this area. Chart no. 31 shows the total hours of sunshine per year in the same locations.

SUMMER

Temperatures in the plateau region vary greatly with elevation. From June into September temperatures usually reach a high in the mid eighties during the day near the Grand Canyon rim, and drop into the mid forties at night. However, in the Canyon itself, the mid summer thermometer usually ranges from 100° to 120°. Remember, in considering temperatures, it is also necessary to examine the elevations.

In the northern plateau area, temperatures in the afternoon reach mid to upper 80s and drop into the comfortable 50s at night. Figures between 100° and 105° have been recorded, but they are rare.

Generally mid to upper 90s can be expected in the mid plateau area, except in places located in the higher regions. Temperatures at night, however, do drop into the sleepable 60s.

South and east of the plateau area is truly a hot spot. Afternoon temperatures generally go over the 100° mark and even though the air is extremely dry the Temperature Humidity Index still reaches the mid to upper 80s, and everyone feels some discomfort.

With increased irrigation projects underway, more moisture is being added to the atmosphere, thus increasing the discomfort. In the Phoe-

Table 84	AVERAGE % OF MAXIMUM POSSIBLE SUNSHINE (% of daylight hours)												
Area Number 10 Southwest	Jan	Feb	March	April	May	June	July	Aug	Sept	Oct	Nov	Dec	Year
Arizona													
Phoenix	76	79	83	88	93	94	84	84	89	88	84	77	85
Yuma	83	87	91	94	97	98	92	91	93	93	90	83	91
New Mexico													
Albuquerque	70	72	72	76	79	84	76	75	81	80	79	70	76
Roswell	69	72	75	77	76	80	76	75	74	74	74	69	74
Texas													
Amarillo	71	71	75	75	75	82	81	81	79	76	76	70	76
Del Rio	53	55	61	63	60	66	75	80	69	66	58	52	63
El Paso	74	77	81	85	87	87	78	78	80	82	80	73	80

PHOENIX, ARIZONA — Elevation 1,109 Feet — Table 85

Month	Average Max	Average Min	Extreme Max	Extreme Min	THI*	Wind Chill Factor*	Precip. Total	Precip. Snow	Not even 0.01" precip.	More than ½" of snow	Clear	Cloudy	Thunder-storms	Fog	90° or higher	32° or lower	% of possible sunshine	Rel. Humidity A.M.	Rel. Humidity P.M.	Wind M.P.H.	Wind Direction	Storm Intensity*
Jan	65	35	85	17		32	1	T	27	0	21	10	T	1			77	69	39	4	E	R-2
Feb	70	39	88	22		37	1	T	24	0	19	9	T	T			79	67	34	5	E	R-2
March	76	44	92	29			1	0	28	0	22	9	T	T			83	61	28	6	E	R-2
April	84	50	104	32	73		T	T	28	0	23	2	1	0			88	51	21	6	E	R-1
May	95	58	113	42	77		T	0	30	0	27	4	1	0			93	42	16	6	E	R-1
June	102	66	117	50	84		T	0	29	0	28	2	1	0			94	37	14	6	E	R-1
July	105	75	118	61	88		1	0	27	0	27	4	6	0			84	53	24	6	E	T-2
Aug	102	74	115	60	87		1	0	26	0	27	4	7	0			84	60	27	6	E	T-2
Sept	98	67	118	49	79		1	0	27	0	27	3	3	0			89	56	27	5	E	T-2
Oct	88	54	104	36	77		T	0	28	0	26	5	1	0			88	56	30	5	E	R-1
Nov	76	41	91	25			T	0	28	0	24	6	T	T			84	64	38	4	E	R-1
Dec	68	37	88	22		34	1	0	27	0	22	9	1	1			77	67	40	4	E	R-2
Year	86	58	118	17			7	T	331	0	293	72	22	2			86	57	28	5	E	

Notes:

T Indicates "trace"

* For full explanation of (T-H-I) "Temperature Humidity Index;" "Wind Chill Factor" and "Storm Intensity," see beginning of Chapter 2.

Average date of first freeze .... December 11

" " last .... January 27

" freeze-free period .... 317 days

10 inches of snow equal approximately one inch of rain.

187

TUCSON, ARIZONA — Elevation 2,584 Feet — Table 86

Month	Average Max	Average Min	Extreme Max	Extreme Min	THI*	Wind Chill Factor*	Precip. Total	Precip. Snow	Not even 0.01" precip.	More than ½" of snow	Clear	Cloudy	Thunder-storms	Fog	90° or higher	32° or lower	% of possible sunshine	Rel. Hum. A.M.	Rel. Hum. P.M.	Wind M.P.H.	Wind Direction	Storm Intensity*
Jan	63	36	87	16		30	1	T	27	0	21	11	0	0			80	40		7	SE	R-2
Feb	67	40	92	20		33	1	T	24	0	19	9	0	0			84	35		7	SE	R-2
March	73	43	92	26			1	T	27	0	21	10	0	0			85	28		8	SE	R-2
April	81	50	102	27	71		T	0	28	0	24	6	1	0			90	22		8	SE	R-1
May	89	57	107	38	79		T	0	30	0	27	4	1	0			93	17		8	SE	R-1
June	98	66	111	47	81		T	0	29	0	28	2	2	0			92	16		8	SSE	R-1
July	99	73	111	63	85		2	0	20	0	22	9	14	0			76	33		8	SE	T-3
Aug	96	72	109	61	83		2	0	22	0	24	7	13	0			80	39		7	SE	T-3
Sept	93	67	107	48	80		1	0	27	0	27	3	5	0			90	29		7	SE	T-2
Oct	84	55	101	35	74		T	T	27	0	27	4	2	0			90	30		7	SE	R-1
Nov	73	44	90	24			1	T	28	0	25	5	0	0			89	30		8	SE	R-2
Dec	66	38	84	18		32	1	T	27	0	22	9	0	0			83	37		7	SE	R-2
Year	82	53	111	16			11	1	316	0	286	79	39	0			86	30		7	SE	

Notes:

T Indicates "trace"

* For full explanation of (T-H-I) "Temperature Humidity Index," "Wind Chill Factor" and "Storm Intensity," see beginning of Chapter 2.

Average date of first freeze .... November 23
"   "   last   "   .... March 6
"   freeze-free period   .... 261 days
10 inches of snow equal approximately one inch of rain.

188

ALBUQUERQUE, NEW MEXICO — Elevation 5,310 Feet — Table 87

Month	Temp Avg Max	Temp Avg Min	Extreme Max	Extreme Min	THI*	Wind Chill Factor*	Precip Total (in)	Precip Snow (in)	Not even 0.01" precip.	More than ½" of snow	Days Clear	Days Cloudy	Thunderstorms	Fog	90° or higher	32° or lower	% of possible sunshine	Rel. Hum. A.M.	Rel. Hum. P.M.	Wind M.P.H.	Wind Direction	Storm Intensity*
Jan	46	22	67	1		11	T	2	27	1	21	10	0	1	0	26	68	51	46	8	N	R, S-1
Feb	52	27	72	−5		15	T	2	24	1	19	9	0	1	0	20	73	43	36	9	N	R, S-1
March	60	32	81	8		19	T	1	27	1	22	9	1	0	0	15	72	33	28	10	NW	R, S-1
April	69	42	88	19			1	1	26	0	21	9	2	0	0	3	76	28	23	11	SE	R, S-1
May	79	52	98	34	70		1	T	27	0	25	6	4	0	2	0	79	26	21	11	SE	T-2
June	89	61	101	45	76		1	0	26	0	27	3	5	0	17	0	84	24	18	10	S	T-2
July	92	66	104	55	80		1	0	22	0	27	4	13	0	23	0	76	34	28	9	SE	T-2
Aug	89	65	100	53	78		1	0	21	0	27	4	13	0	17	0	75	38	32	8	SE	T-2
Sept	82	58	98	39	73		1	T	25	0	26	4	4	0	6	0	82	34	28	9	SE	T-2
Oct	71	45	87	26			1	T	26	0	25	6	3	0	0	0	79	36	32	8	SE	R-1
Nov	57	31	74	10		21	T	2	28	1	25	5	1	1	0	17	78	41	35	8	N	R, S-1
Dec	47	25	68	7		17	1	2	27	1	22	9	0	1	0	26	70	50	46	8	N	R, S-1
Year	69	44	104	−5			9	9	307	5	187	78	46	4	65	107	76	37	31	9	SE	

Notes:

T Indicates "trace"

* For full explanation of (T-H-I) "Temperature Humidity Index," "Wind Chill Factor" and "Storm Intensity," see beginning of Chapter 2.

Average date of first freeze .... October 29
    "   "   last   "   .... April 16
    "   freeze-free period .... 196 days
10 inches of snow equal approximately one inch of rain.

AMARILLO, TEXAS

Elevation 3,590 Feet

Table 88

Month	Temperatures Average Max.	Average Min.	Extreme Max.	Extreme Min.	T H I*	Wind Chill Factor*	Precip. Total	Snow	Not even 0.01" precip.	More than ½" of snow	Clear	Cloudy	Thunder-storms	Fog	90° or higher	32° or lower	% of possible sunshine	Rel. Hum. A.M.	Rel. Hum. P.M.	Wind M.P.H.	Wind Direction	Storm Intensity*
Jan	50	24	79	-9		1	1	4	27	1	19	12	0	3	0	27	70	69	47	13	SW	R, S-2
Feb	55	28	88	3		4	1	3	24	1	17	11	0	4	0	23	68	73	50	14	SW	R, S-2
March	62	33	90	7		11	1	2	27	0	21	10	1	3	0	15	72	65	37	16	SW	R, S-2
April	72	43	98	25			1	T	24	0	20	10	3	2	1	2	71	64	35	15	SW	R-1
May	80	53	99	33	72		3	T	22	0	21	10	8	2	8	0	71	70	39	15	S	T-3
June	91	63	104	43	79		3	0	22	0	25	5	9	1	11	0	75	80	48	14	S	T-3
July	94	67	102	55	81		2	0	22	0	26	5	10	1	23	0	77	74	43	12	S	T-3
Aug	93	66	104	52	80		3	0	23	0	26	5	9	1	17	0	77	76	46	12	S	T-3
Sept	85	58	98	39	75		2	T	25	0	24	6	4	2	6	0	76	82	50	13	S	T-2
Oct	75	46	94	30			2	T	26	0	23	8	2	2	2	0	75	67	35	13	SW	R-2
Nov	60	32	82	15		14	1	2	27	0	24	6	1	2	0	11	73	73	47	13	SW	R-1
Dec	52	26	76	-3		4	1	2	27	1	21	10	0	2	0	26	68	70	49	13	SW	R, S-2
Year	73	45	104	-9			20	13	296	3	267	98	48	24	69	104	73	72	44	14	SW	

Notes:

T Indicates "trace"

* For full explanation of (T-H-I) "Temperature Humidity Index," "Wind Chill Factor" and "Storm Intensity," see beginning of Chapter 2.

Average date of first freeze .... October 23
" " last .... April 20
" freeze-free period .... 185 days
10 inches of snow equal approximately one inch of rain.

EL PASO, TEXAS — Elevation 3,920 Feet — Table 89

Month	Average Max.	Average Min.	Extreme Max.	Extreme Min.	T-H-I*	Wind Chill Factor*	Precip. Total	Precip. Snow	Not even 0.01" precip.	More than ½" of snow	Clear	Cloudy	Thunder-storms	Fog	90° or higher	32° or lower	% of possible sunshine	Rel. Hum. A.M.	Rel. Hum. P.M.	Wind M.P.H.	Wind Direction	Storm Intensity*
Jan	56	30	75	-8		17	T	1	28	1	21	10	0	1	0	21	77	43	33	10	N	R, S-1
Feb	62	36	79	11		23	T	1	25	0	21	7	0	0	0	13	81	36	26	11	N	R, S-1
March	69	40	88	15			T	T	29	0	22	9	1	0	0	5	83	31	22	12	WSW	R, S-1
April	78	49	98	33	68		T	T	28	0	24	6	1	0	2	0	86	22	14	12	WSW	R-1
May	87	57	100	31	72		T	0	29	0	27	4	2	0	15	0	89	21	13	12	WSW	R-1
June	95	67	106	51	83		1	0	26	0	27	3	5	0	26	0	88	26	18	11	S	T-2
July	95	69	106	64	82		1		23	0	25	6	10	0	28	0	79	41	30	9	SSE	T-2
Aug	93	68	102	58	81		1		24	0	27	4	11	0	23	0	80	42	33	9	S	T-2
Sept	88	61	98	45	77		1		26	0	25	5	4	0	7	0	83	47	36	9	S	T-2
Oct	79	50	92	27	70		1		27	0	26	5	2	0	1	0	85	32	24	8	N	T-2
Nov	66	36	81	23		26	T	1	28	0	25	5	0	0	0	6	82	40	33	9	N	R-1
Dec	58	31	73	11		20	T	1	28	0	23	8	0	0	0	19	77	46	37	9	N	R, S-1
Year	77	49	106	-8			8	4	321	1	293	72	36	1	101	64	83	36	27	10	N	

Notes:

T Indicates "trace"

* For full explanation of (T-H-I) "Temperature Humidity Index," "Wind Chill Factor" and "Storm Intensity," see beginning of Chapter 2.

Average date of first freeze .... November 11
    "   "   last   "   .... March 13
    "   freeze-free period .... 243 days

10 inches of snow equal approximately one inch of rain.

nix area readings as high as 118° have been recorded. At night expect the temperatures to drop only into the mid 70s; still too warm for comfortable sleeping. Air conditioning in both houses and cars is almost essential during the summer months.

Further east (El Paso), temperatures reach the mid 90s on most days but the increased moisture coming from the Gulf of Mexico brings the discomfort index to a point where almost everyone feels quite uncomfortable. During heat waves temperatures over 100° are common. Because of the moisture, temperatures at night remain in the upper 60s to near 70°. Still somewhat uncomfortable for sleeping.

In the panhandle area of Texas, temperatures in the low 90s are common along with high humidities, creating somewhat uncomfortable weather through the summer months. Precipitation is heaviest in summer and occurs mostly in areas of high terrain. Thunderstorms during the afternoon hours are responsible for just about all of the summer rainfall. A general rule is that afternoon thunderstorms will occur in the higher country about every second or third day, and every four or five days in the lower regions. These storms are usually brief and only temporarily dampen any outdoor activity.

THE PACIFIC NORTHWEST—Area Number 11

See Charts Nos. 32 and 33

(Washington and Oregon)

The spectacular scenery coupled with the wide variations in climate and topography are a most pleasant surprise to many first time visitors to the Pacific northwest.

This somewhat isolated outdoor wonderland is all too often written off as the land of the web-foot. To be sure, you can find plenty of rain, particularly during the winter months along the Pacific coast and in the area of Puget Sound. The rain forest, on the Olympic Peninsula, is drenched with 140 inches a year but Sequim, only sixty miles away in the lee of the mountains on the "suncoast," averages only 16 to 20 inches. Ellensburg, in the drier interior, manages with a total annual precipitation of nine inches.

But in addition to the mountain ranges, the Pacific Northwest can be divided into two broad regions. West of the Cascades, both Washington and Oregon enjoy a typical marine westcoast climate, while most of the country to the east experiences a more normal continental climate.

The former is characterized by cool winters and agreeably mild to warm summers. Occasionally heat waves invade the coastal strip, but temperatures are generally pleasant. Although precipitation occurs every month throughout the year, winter is the season of the heaviest rainfall. It's also a period of almost continuous fog or overcast skies. Due to the moderating effect of the prevailing air currents off the Pacific, there are seldom very severe hot or cold spells.

The much larger land area east of the Cascades experiences far greater temperature extremes. Readings of 105° to −20° are not uncommon. In contrast to the coast, this region gets rather limited rainfall. The prevailing, moisture-laden air from the Pacific is wrung out and it's not uncommon to see in viewing these mountains from the air lush green on the western side while the opposite slopes appear dry and brownish. The west side probably averages 100 inches of precipitation a year, which may be ten to twenty times as great as the fall on the eastern slopes.

SUMMER

Summer is a most delightful season along the coast. Due to the persistent breezes from the Pacific Ocean, temperatures are almost always agreeable. Daytime highs in the state of Washington are usually between 70 and 75 degrees, while night time temperatures in the 50s are comfortable for sleeping. Conditions along the Oregon coast are very similar, with maximum afternoon temperatures ranging in the low 80s and at night dipping into the upper 50s.

In the Puget Sound area figures are slightly higher, reaching 85–90 degrees during the afternoons. Extremely hot weather, however, is practically unknown around Puget Sound, 95 degrees being the highest ever recorded.

Occasionally the cool Pacific breezes fail and the coastal sections of Washington and Oregon experience unusually warm weather. During these rare heat waves, temperatures have gone into the high nineties along the coast of Washington and over 100° along the Oregon coast.

The summer climate in the coastal northwest is said to be one of the healthiest in the world. If it weren't for the damp, dreary winters this

PACIFIC
NORTHWEST

AREA No. 11   CHART No. 32
APRIL 1ST To OCTOBER 1ST

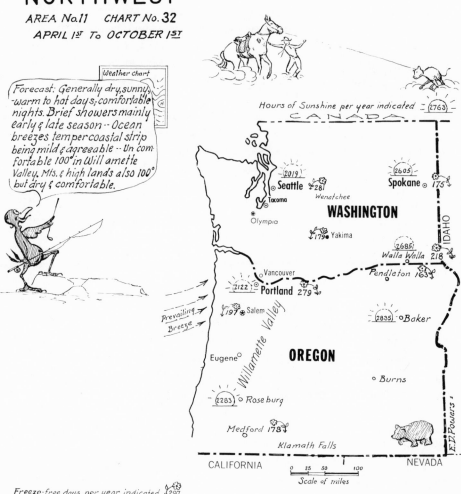

Weather chart

Forecast: *Generally dry, sunny, warm to hot days; comfortable nights. Brief showers mainly early & late season -- Ocean breezes temper coastal strip being mild & agreeable -- Uncomfortable 100° in Willamette Valley. Mts. & high lands also 100° but dry & comfortable.*

Hours of Sunshine per year indicated [2763]

CANADA

[2019] Seattle
Tacoma
Wenatchee
Olympia
WASHINGTON
[2605] Spokane 175
179 Yakima

IDAHO
[2685]
Walla Walla 218
Pendleton 163

Vancouver
[2122] Portland 279

197 Salem
Willamette Valley

[2835] Baker

Eugene

OREGON

Burns

[2283] Roseburg

Medford 178

Klamath Falls

E.D. Powers

CALIFORNIA   NEVADA

0   25   50   100
Scale of miles

Prevailing Breeze

Freeze-free days per year indicated 297

E.D. Powers

PACIFIC NORTHWEST

AREA No.11   CHART No.33

OCTOBER 1ST to APRIL 1ST

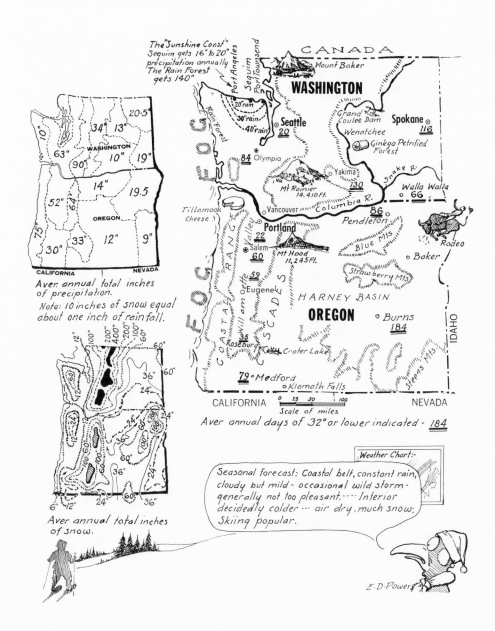

The "Sunshine Coast" Sequim gets 16" to 20" precipitation annually. The Rain Forest gets 140"

Aver. annual total inches of precipitation.

Note: 10 inches of snow equal about one inch of rainfall.

Aver annual total inches of snow.

Aver annual days of 32° or lower indicated - 184

Scale of miles

Weather Chart:-

Seasonal forecast: Coastal belt, constant rain, cloudy but mild - occasional wild storm - generally not too pleasant.---- Interior decidedly colder --- air dry, much snow; Skiing popular.

E·D·Powers

area could well become a real mecca for retirees. Those fortunates who can migrate with the swallows are settling there in increasing numbers.

The Willamette Valley of Oregon presents another exception to the generally comfortable temperatures west of the Cascades. This large fertile stretch, the market basket of much of the state, lies between the Coast Range on the west and the Cascades on the east. The former block the cool sea breezes, resulting in temperatures consistently reaching 100°.

To designate the Pacific Northwest as the rainiest region in the United States is very misleading. Actually, a day of rain during the months of June through September is the exception. Showers occur only about four times in July and then dropping less than an inch of rain. Amounts and frequencies are slightly higher during the other summer months, but still remain insignificant. In the Olympic Mountain area precipitation is more frequent but again, mainly in the form of showers, totaling approximately eight inches during the summer months.

The lack of summer rainfall is welcomed by the tourists as is the abundant sunshine. The sun is visible, through clear to partly cloudy skies, approximately eight hours per day along the northwest coast. Further inland, in the shadow of the coastal ranges of Oregon, the sun shines nine hours or more per day. July is the sunniest month in the northwest, the sun being on the job 60–70% of the time from sunrise to sunset. Along the immediate coast, clouds are more prevalent but still the sun is visible for about 50% of the daylight hours.

The air in the Northwest is usually clear, with visibility good to excellent during the summer months. Thunderstorms are almost non-existent along the coast. Further inland, in the Puget Sound area, thunderstorm occurrence increases to less than five during the summer months. Hail storms follow about the same pattern of occurrence.

Cascade Mountains and East

As we enter the continental climate east of the Cascades, very warm to hot daytime temperatures can be expected. Even in the Cascades, temperatures reach into the middle 90s on most summer days. It must be noted, however, that temperatures within the mountains always vary with elevation: as altitude increases, temperatures decrease. So as one approaches the numerous higher peaks such as Mount Adams, Mount

Rainier and Mount Hood, temperatures are low enough to maintain perpetual snow. Further east, in the Okanogan Highlands as far as Spokane, temperatures near 100° are common. Spokane's temperatures, however, are generally in the eighties. South of the highlands on the Columbia plateau of Washington and Oregon, including Pasco, Walla Walla, Yakima and Pendleton, daytime temperatures approaching 100° are the rule. Highs of 115° have been experienced on the plateau.

Because the air is so dry these extreme temperatures can be tolerated, although some people are a bit uncomfortable. During the night, again due to the low himidity, temperatures drop into the comfortable 50s throughout the entire area.

The Cascade mountains block the moisture from entering this whole region and as a result practically no precipitation occurs throughout the summer. During July and August, there is less than one inch of rain per month, which occurs mainly in the form of an occasional thundershower during the hot afternoons. Rain rarely is experienced on more than four days per month during July and August in any of the areas east of the Cascade mountains. In the Cascade mountain area itself, rainfall is slightly more abundant but not enough to cause much inconvenience.

The Cascades in summer are sunny and pleasant. The sun shines ten to eleven hours per day, which is about 70% of the possible maximum. Further to the east, in the Highlands and Plateau, twelve to thirteen hours of sunshine is common. Visibility is usually excellent during the summer months, affording a fine opportunity to view the awesome peaks of the Cascade ranges.

Table 90 lists the percent of maximum possible sunshine in a number of locations throughout the two states. Chart no. 32 also shows the total number of hours of sunshine per year in six such spots.

Coastal Areas—West of Cascades

Winter

Just as the equable temperatures of the west coast marine climate insure cool weather along the coastal sections in summer, they also provide mild temperatures throughout the winter. The daytime temperatures reach into the mid to upper forties and at night drop into the

197

thirties. Brisk damp winds from the ocean cause these deceptively mild temperatures to feel cold and uncomfortable. In the Puget Sound area the thermometer usually drops into the 20s at night while during the day it climbs to the low to mid forties. In Seattle, temperatures have dropped to near zero on occasion. In the Olympic mountains, because of their increased altitude, the temperatures generally dip to the teens at night while daytime readings remain in the thirties.

The southern coast of Oregon is somewhat warmer than further north. Daytime highs in Portland, which is situated in the northern end of the Willamette Valley, reach the mid to upper 40s and drop to the mid to upper 30s at night. Occasionally, the mercury has dipped to the single numbers, but this is rare. Portland is slightly milder than the rest of the valley because of its location on the Columbia River. The river valley allows the tempering Pacific breezes to flow into Portland, but the Coastal Ranges of Oregon block these breezes from affecting other portions of the Willamette Valley. In the valley, therefore, 10° temperatures at night are not unusual; daytime temperatures are generally found approaching 40°

Although temperatures do not prohibit travel during the winter, the precipitation is certainly a deterring factor. Precipitation in the coastal sections of the Pacific northwest is very heavy and consistent. During the months of December through February, as much as 30–40 inches fall along the immediate Washington coast, increasing to about 50 inches in the Olympic Sound area. Generally, these mountains receive about 20 inches of snow due to their height, but directly along the coast all precipitation is in the form of rain. Further to the east, around the Puget Sound area, only about 10 inches fall through the winter months. Snowfall in this area averages between 10–15 inches, with snow remaining on the ground about 10 to 20 days during the winter. The lower amount of precipitation around Puget Sound is caused by the Olympic mountains which block the moisture from the Pacific Ocean. As one travels east from Puget Sound toward the Cascades, and out of the Olympic rain shadow the precipitation increases and the amount of snowfall increases rapidly.

Not only is the rainfall a factor in discouraging travel in these areas during the winter months but so too are the visibility and sky conditions. All of southwest Washington, from the coast to the Cascades, can expect very poor visibility and low hanging clouds four out of every ten days. The Olympic peninsula and all of Vancouver Island except the southern tip can expect about three out of ten days to have these same unfavorable conditions. Again the Puget Sound area is exceptional. It and the southern tip of Vancouver Island experience poor visibility and low clouds only one to two days out of ten.

Conditions are quite similar in Oregon. Along the northern coast about 40 inches of precipitation occurs, while the southern coast receives about 20–25 inches from December through February. In the Willamette Valley, one can expect approximately 20 inches during the same period. Unlike the Washington coast, the Oregon coast will average about 10 inches of snow during the winter months with snow remaining on the ground only about ten days through the entire winter. In the Willamette Valley snowfall averages about 20 inches. The amount of snow steadily increases as you move eastward toward the Cascades. Portland, however, usually receives only about 10 inches of snow per winter. Visibility, with low clouds, presents a problem in western Oregon as it does in Washington. On the extreme north coast of Oregon four of every ten days during the winter have a low overcast sky with poor visibility. The rest of the coast, including the Willamette Valley, experiences about three days in ten with the same conditions. Traveling eastward toward the Cascades, these conditions improve somewhat. However, one can expect only 2 or 3 hours of sunshine per day during the winter throughout the entire area.

Winter travel in the Pacific northwest, west of the Cascades, therefore is discouraged by the consistent pattern of rainfall and occasional snow, compounded by numerous days with overcast skies and poor visibility.

The Pacific Northwest—East of Cascades and Cascades

WINTER

In marked contrast to the rather mild but damp temperatures of the coastal areas, interior Washington and Oregon in the Cascades and east are decidedly colder, especially since they are often subjected to cold arctic air masses from the north.

Average winter temperatures in the Cascades are approximately 30 degrees, with cold spells plunging the mercury to −10°. Mount Rainier and other high peaks have recorded readings as low as −30°. The regions east of the Cascades in both Washington and Oregon have similar temperatures throughout, with some variations in the Columbia river valley and in the higher elevations. The average January temperature (January is the coldest month) is 25°. Spokane's daytime temperatures are approximately 30° while temperatures at night drop to 20°. This small difference in range from the maximum during the day to the minimum at night is caused by the almost continuous presence of clouds. December and February are somewhat milder. Occasionally arctic outbreaks of cold air can drop the thermometer below zero. The Columbia river valley, marking the boundary between Washington and Oregon, is an area of milder temperatures. This is due to the marine influence which penetrates the mountain gap and tempers the valley weather almost to the Idaho border. For example, in Pendleton, near the Columbia River, January daytime temperatures are in the high 30s. At night they drop into the mid 20s. On the other hand, the numerous mountain regions record much colder temperatures, while nearby areas will be 10° lower than the valley. This is a good example of how topography can so radically affect the climate in a local area and why overall weather descriptions can be quite misleading.

Winter precipitation in the Cascades falls mainly as snow. It may accumulate to over 100 inches throughout the winter. In the rain shadow, directly east of the mountains, precipitation decreases sharply to 4 inches from December through February and remains relatively low throughout eastern Washington and Oregon. Spokane receives the majority of its precipitation in the form of snow. Pendleton, which is cow country and famous for its rodeo, has little snow because it is warmer (see tabulation). Although the amounts of precipitation are low in this area, the skies are nevertheless very overcast. Washington and northern Oregon have an average of 2½ to 3 hours of sunshine per day from December thru February. Central and southern Oregon, boasting sunnier skies, receive an average of four to five hours per day. While the coast is having the miseries, this whole high country is enjoying fine skiing and other winter activities.

The Pacific Northwest—West of Cascades

Spring and Fall

Spring and fall are transition periods. Daytime temperatures during both seasons are in the upper 50s and low 60s dropping to the mid 40s at night. In the transition, however, each month's figures vary. March, in early spring, corresponds with November of late fall with daytime temperatures in the mid 50s and nights in the low 40s. These are typical temperatures in Portland, the City of Roses, which is characteristic of coastal Oregon. Washington, in the Puget Sound area, is slightly cooler. Seattle temperatures in March and November just approach 50° during the day and drop to the upper 30s at night. Readings immediately along coastal Washington do not deviate significantly from the Seattle figures. April and October in Portland have highs in the mid 60s and low 70s and lows in the mid 50s during the night. Again Seattle remains consistently cooler by a few degrees throughout these months.

The entire section west of the Cascades has more abundant rainfall during autumn than spring. In both seasons greater amounts fall on the northern coast. In autumn, for example, the Washington coast receives 35 inches, which decreases gradually to 15 inches on the southern Oregon coast. In the spring, coastal Washington averages 25 inches of rain again diminishing slightly in Oregon. The Willamette Valley in Oregon, in the lee of the coastal ranges averages about five inches less than the coast during both seasons.

A most remarkable exception however is the Puget Sound area, where the coast receives 25 to 35 inches of precipitation. Just a short distance to the east (about 60 miles) the amount of precipitation drops suddenly to about 16 inches per year. The Puget Sound coast from Port Angeles to Port Townsend, known as the Sunshine Belt, does not exceed six inches in either spring or fall. Farther south in Seattle, rainfall increases to about 8″ in spring and 10″ in fall.

Traveling along the coast of Washington and Oregon in autumn one can expect one in every four days to be overcast, whereas in the spring only one in ten days is overcast and foggy. Again Puget Sound around Port Angeles and Port Townsend demonstrates a notable exception. They experience overcast skies only one day in every eight during autumn and less than one in ten during spring. Considering the pleasant temperatures and the abundance of clear sunny skies, the Puget Sound is a fine area to visit and live in—not only in spring and fall but throughout the year.

The Pacific Northwest—Cascades and East of Cascades

Spring and Fall

Since spring and fall lie between the two extremes in temperature they show characteristics of both winter and summer but in a modified form. The months of March, April, October and November can be especially unpredictable. For example, in Pendleton during the month of March the figures have climbed as high as 78° and dropped as low as 10°. Spokane, in November, can record highs in the 60s and within the same month a low of −11°. In general, however, the eastern sections of Washington and Oregon have daytime temperatures ranging from the 40s in March to the mid 60s in May, and nights usually from the low 30s to mid 40s. The temperature spread from September through November is somewhat greater. Average daytime readings in September are in the low 70s but by November they have already dropped to the low 40s.

At night, September readings are in the upper 40s and gradually decrease to just below freezing in November. Pendleton and other sections along the Columbia River are slightly milder. It must be noted, however, that average figures can be misleading in this area, particularly in the spring and fall when rapid and extreme changes can occur. This is good to remember in regard to statistics related to almost any facet of weather—be it temperatures, precipitation or rainbows—and applies to averages covering large areas or long periods of time.

Table 90	AVERAGE % OF MAXIMUM POSSIBLE SUNSHINE (% of daylight hours)												
Area Number 11 Pacific Northwest	Jan	Feb	March	April	May	June	July	Aug	Sept	Oct	Nov	Dec	Year
Washington													
Seattle	27	34	42	48	53	48	62	56	53	36	28	24	45
Spokane	26	41	53	63	64	68	82	79	68	53	28	22	58
Walla Walla	24	35	51	63	67	72	86	84	72	59	33	20	60
Yakima	34	49	62	70	72	74	86	86	74	61	38	29	65
Oregon													
Baker	41	49	56	61	63	67	83	81	74	62	46	37	60
Portland	27	34	41	49	52	55	70	65	55	42	28	23	48
Roseburg	24	32	40	51	57	59	79	77	68	42	28	18	51

SEATTLE, WASHINGTON — Elevation 386 Feet — Table 91

Month	Avg Max.	Avg Min.	Extreme Max.	Extreme Min.	THI*	Wind Chill Factor*	Precip. Total	Precip. Snow	Not even 0.01" precip.	More than ½" of snow	Clear	Cloudy	Thunderstorms	Fog	90° or higher	32° or lower	% of possible sunshine	R.H. A.M.	R.H. P.M.	Wind M.P.H.	Wind Direction	Storm Intensity*
Jan	46	37	66	11		16	5	4	11	2	7	24	0	5	0	7	28	86	79	11	SE	R, S-3
Feb	49	38	70	12		20	4	1	12	1	6	22	0	4	0	3	34	85	73	11	SW	R, S-3
March	53	40	75	22		23	3	1	13	1	8	23	0	3	0	1	42	85	65	12	SW	R, S-2
April	59	44	87	31		27	2	T	16	0	10	20	0	1	0	0	47	85	58	11	SW	R-2
May	66	49	92	35			2	0	21	0	14	17	1	1	0	0	52	85	56	11	SW	R-2
June	70	53	100	45			1	0	19	0	12	18	1	1	0	0	49	84	65	10	SW	R-2
July	75	56	100	48	69		1	0	26	0	22	9	1	3	1	0	63	85	51	10	SW	R-2
Aug	74	56	97	48	68		1	0	24	0	18	13	1	4	0	0	56	87	54	9	SW	R-2
Sept	69	53	92	42			2	0	21	0	16	14	1	8	0	0	53	89	61	9	N	R-2
Oct	60	48	78	30			3	0	16	0	11	20	1	9	0	0	37	90	73	10	S	R-2
Nov	52	42	70	13		26	5	1	11	0	7	23	0	7	0	1	28	88	80	10	S	R, S-3
Dec	48	40	65	11		20	5	1	11	1	5	26	0	7	0	3	23	87	81	11	SSW	R, S-3
Year	60	46	100	11			34	8	202	5	136	229	6	53	1	15	45	86	65	11	SW	

Notes:

T Indicates "trace"

* For full explanation of (T-H-I) "Temperature Humidity Index," "Wind Chill Factor" and "Storm Intensity," see beginning of Chapter 2.

Average date of first freeze .... December 1
" " last " ....February 23
" freeze-free period .... 281 days
10 inches of snow equal approximately one inch of rain.

SPOKANE, WASHINGTON — Elevation 2,357 Feet — Table 92

Month	Temperatures Average Max.	Average Min.	Extreme Max.	Extreme Min.	T-H-I*	Wind Chill Factor*	Precip. Total	Precip. Snow	Not even 0.01" precip.	More than ½" of snow	Clear	Cloudy	Thunder-storms	Fog	90° or higher	32° or lower	% of possible sunshine	Rel. Hum. A.M.	Rel. Hum. P.M.	Wind M.P.H.	Wind Direction	Storm Intensity*
Jan	31	19	49	−13		9	2	19	14	6	8	23	0	9	0	28	28	80	74	8	NE	S, R-3
Feb	37	23	56	1		10	2	9	16	4	8	20	0	7	0	22	39	78	69	9	SSW	S, R-2
March	47	29	71	2		18	2	6	19	2	12	19	0	3	0	21	53	70	57	10	SSW	R, S-2
April	59	36	80	24		27	1	1	22	0	14	16	1	1	0	10	62	56	42	10	SW	R-2
May	69	43	87	30			1	T	21	0	16	15	1	1	0	2	62	55	44	8	SSW	R-2
June	75	49	97	34			2	T	22	0	17	13	3	1	3	0	67	47	33	9	SSW	R-2
July	86	55	102	39	72		T	0	27	0	27	4	2	0	10	0	82	37	33	8	SW	T-2
Aug	83	53	108	35	72		T	0	26	0	24	7	3	0	7	0	77	45	30	8	SW	T-2
Sept	75	47	93	32			1	0	24	0	20	10	1	1	1	0	72	47	31	8	NE	R-2
Oct	60	38	85	24		31	2	1	22	0	15	16	0	5	0	8	51	70	53	8	SSW	R-2
Nov	43	29	58	−2		21	2	6	18	2	10	20	0	8	0	20	30	83	75	8	NE	R, S-2
Dec	36	24	51	−20		14	2	17	15	5	6	25	0	13	0	27	21	86	83	8	NE	R, S-3
Year	58	37	108	−20			17	58	247	19	177	188	12	47	20	138	58	63	51	8	SSW	

Average date of first freeze .... October 12
" " last " .... April 20
" freeze-free period .... 175 days
10 inches of snow equal approximately one inch of rain.

Notes:
T Indicates "trace"
* For full explanation of (T-H-I) "Temperature Humidity Index;" "Wind Chill Factor" and "Storm Intensity," see beginning of Chapter 2.

PORTLAND, OREGON Elevation 30 Feet Table 93

Month	Average Max.	Average Min.	Extreme Max.	Extreme Min.	THI*	Wind Chill Factor*	Precip. Total	Precip. Snow	Not even 0.01" precip.	More than ½" of snow	Clear	Cloudy	Thunderstorms	Fog	90° or higher	32° or lower	% of possible sunshine	Rel. Hum. A.M.	Rel. Hum. P.M.	Wind M.P.H.	Wind Direction	Storm Intensity*
Jan	44	33	62	−2		23	5	5	11	2	5	26	0	4	0	14	24	82	77	10	ESE	R, S-3
Feb	49	35	66	−3		28	4	1	12	1	6	22	0	3	0	9	32	80	69	9	ESE	R, S-3
March	54	38	80	19			4	1	14	0	6	25	0	2	0	6	37	71	61	9	ESE	R, S-3
April	62	42	87	29			2	T	16	0	10	20	0	1	0	1	47	68	55	7	NW	R-2
May	68	47	92	29			2	T	19	0	11	20	2	0	0	0	51	66	54	7	NW	R-2
June	72	52	100	39			2	0	21	0	12	18	1	0	1	0	47	65	49	7	NW	R-2
July	79	56	107	43	73		T	0	28	0	23	8	1	0	3	0	67	63	46	7	NW	R-1
Aug	78	55	100	44	73		1	0	27	0	21	10	1	0	3	0	61	66	48	7	NW	R-1
Sept	74	51	101	34			2	T	22	0	18	12	1	4	1	0	58	67	50	6	NW	R-2
Oct	63	45	89	26			4	T	19	0	12	19	0	8	0	1	38	80	66	6	ESE	R-3
Nov	52	38	69	13			5	T	13	0	8	22	0	6	0	6	29	82	75	8	ESE	R-3
Dec	47	36	64	6		26	6	1	12	0	4	27	0	5	0	9	21	84	79	10	ESE	R-3
Year	62	44	107	−3			37	8	214	3	136	229	7	33	8	45	45	73	61	8	NW	R, S-3

Notes:

T Indicates "trace"

* For full explanation of (T-H-I) "Temperature Humidity Index," "Wind Chill Factor" and "Storm Intensity," see beginning of Chapter 2.

Average date of first freeze .... November 28
" " " last " ....March 11
" " freeze-free period .... 257 days
10 inches of snow equal approximately one inch of rain.

205

PENDLETON, OREGON — Elevation 1,493 Feet — Table 94

Month	Avg Max	Avg Min	Extreme Max	Extreme Min	T H I*	Wind Chill Factor*	Precip Total	Snow	Not even 0.01" precip	More than ½" of snow	Clear	Cloudy	Thunderstorms	Fog	90° or higher	32° or lower	% of possible sunshine	Rel. Hum. A.M.	Rel. Hum. P.M.	Wind M.P.H.	Direction	Storm Intensity*
Jan	39	25	67	−22		14	1	8	18	3	9	22	0	7	0	22	30	77	76	8	SE	S, R-2
Feb	45	29	66	−18		17	1	4	17	2	8	20	0	5	0	16	40	71	66	9	SE	R, S-2
March	54	36	79	10		24	1	1	20	0	13	18	0	2	0	10	51	58	49	10	W	R, S-2
April	64	41	89	18			T	T	21	0	16	14	1	0	0	2	62	51	41	11	W	R-2
May	72	47	99	25			1	T	23	0	19	12	2	0	1	0	68	47	37	10	W	R-2
June	79	53	108	36	70		1	0	23	0	20	10	2	0	4	0	70	43	32	11	W	R-2
July	89	58	110	42	76		T	0	29	0	28	3	2	0	14	0	85	34	23	10	WNW	T-1
Aug	87	57	113	41	75		T	0	28	0	27	4	2	0	10	0	83	37	26	9	SE	T-1
Sept	78	50	102	30	68		1	0	26	0	23	7	1	0	3	0	75	43	32	9	SE	R-2
Oct	65	43	86	11			1	T	23	0	18	13	0	1	0	2	60	57	50	8	SE	R-2
Nov	49	33	74	−6		27	1	1	19	0	12	18	0	6	0	13	38	72	71	8	SE	R, S-2
Dec	43	30	67	−12		19	2	3	18	1	7	24	0	8	0	18	20	79	80	8	SE	R, S-2
Year	64	42	113	−22			12	17	264	6	200	165	11	30	31	84	61	56	49	9	SE	

Average date of first freeze .... November 28
" " " last " .... March 1
" freeze-free period .... 209 days
10 inches of snow equal approximately one inch of rain.

Notes:
T Indicates "trace"
* For full explanation of (T-H-I) "Temperature Humidity Index," "Wind Chill Factor" and "Storm Intensity," see beginning of Chapter 2.

Although the temperatures between the Cascade Ranges and the land east do not vary significantly, there is a marked difference in rainfall. The Cascades receive about 10–15 inches in the spring with the greater amounts falling in the north. The fall precipitation ranges from 25 inches in the north to 15 inches in the south. The area to the east of the mountains has an average of 2 to 4 inches in the spring and slightly more in autumn. The months of March and November show many winter characteristics. Although rainfall in November averages only 2 inches in Spokane, for example, about 20 days of this month are overcast. March has similiar statistics on cloud cover but with only one inch of rain. The remaining months of spring and fall approach summer conditions with decreasing cloudiness. September especially is a pleasant month with agreeable temperatures, low humidities, little rain and an abundance of sunny skies.

Although it can be said without exaggeration that this is truly a sportsman's paradise, and a joy to the outdoor enthusiast, the Pacific northwest is not nearly as popular with either tourist or retiree as might be expected. Part of the explanation may be that it is somewhat off the beaten track. Perhaps its inaccurate and common description as the wettest place in the United States may also be a factor. As we mentioned before, there is a wide choice of wet or dryness,—from the 140-inch annual deluge in the rain forest to the 5 or 10-inch total per year east of the Cascades.

The coastal strip, including the Puget Sound environs, with its glowing, ruddy-complexioned inhabitants, reminds one of the British Isles. This area enjoys the mildest climate of any in comparable latitudes across the entire country—particularly in summer—and is one of the healthiest places in the world.

Perhaps the weather story in this chapter will tempt more people to venture into this largely unspoiled country. The unique Oregon coast with its 36 state parks linked down the 360 miles of magnificent seascape—all but 23 miles of which have been set aside for public use— is in itself more than adequate reward for undertaking the trek.

CALIFORNIA—Area Number 12

See Charts Nos. 34 and 35

California offers a wide range of weather conditions. Although most people divide the state as northern and southern California, meteorologists classify it in terms of at least six climatic zones.

This is hardly surprising. You can find quite a few climate conditions in the 775 miles between San Diego and Crescent City—that is about the same distance as from Chicago to Charleston, or London to Vienna.

Summer along the coastline features cool nights, ocean breezes and early morning fogs. Virtually no snow falls in this narrow plain. The thermometer rarely drops to the freezing point and the annual rainfall, (occurring mostly in the late fall, winter and spring), ranges from an average of 10 inches at San Diego to 22 inches at San Francisco. The mountain barriers, made up of the Sierra Nevadas, Coast Range and the Cascades extending the full length of the state, coupled with the prevailing Pacific breezes, insure a lack of uniformity inland.

Table 95 lists the percentage of maximum possible sunshine in a number of locations stretching the length of the state. See chart no. 35, which shows the total number of hours sunshine per year in seven such spots.

We are treating it in six areas, as follows:

1. *The northern coast* is a region of heavy rainfall, frequent fogs and even temperatures. This stretch extends from the Oregon border, south to San Francisco. Eureka, in the midst of the redwood country, is the only large city in this section.

2. *The mid-coast* from San Francisco to Pismo Beach, which includes Oakland, Santa Cruz, Monterey, and Morro Bay—an area of mild winters, cool, sunny and dry summers with fresh ocean breezes and heavy early morning fog.

3. *The southern coast,* spanning from Santa Barbara, Los Angeles, Long Beach, Laguna Beach, Del Mar and San Diego—a sloping coastal plain where the climate is tempered by the ocean.

4. *Above the 2000 foot level* in the Sierra Nevada and parts of the Coast Range is a zone of winter snows, moderate rainfall and marked fluctuations in temperature. Mount Shasta, Redding, Yreka, Truckee and Bishop are all in this elevated area.

5. *The Great Central Valley,* which actually includes two valleys, is a vast flat, farm land, extending 460 miles north and south. The lower two-thirds of this area, the San Joaquin Valley, gets meager rainfall while the Sacramento Valley, in the north, gets a moderate 15 to 20 inches annual precipitation. Greater extremes of temperature occur

here in both winter and summer than along the coast. Red Bluff, Sacramento, Merced, Fresno, Hanford, Chico and Bakersfield are strung the length of this bountiful agricultural land.

6. *The Mojave–Colorado Desert area* is a land of sparse rainfall, extreme summer heat, unusually mild winters in the low Imperial and Coachella Valley regions but severely cold at the higher elevations.

We will have to treat each climate region separately:

The North Coast, Oregon Border to San Francisco

WINTER

Winters in this area are mild with very little variation in temperature. Afternoon readings will reach the mid 50s and drop into the low 40s at night.

This section of the coast is characterized by cloudy weather with frequent rains, especially during late fall, winter and early spring. Rainfalls are occasionally heavy as Pacific storms travel inland. Count on skies being cloudy 60–70% of the time.

SPRING AND FALL

Afternoon spring temperatures generally reach the upper 50s to near 60°, then drop back into the cool mid to upper 40s in the evening. Fall figures are somewhat higher, being in the low 60s during the afternoon and dropping to the low 50s at night. Clouds prevent any substantial variation in temperature.

Spring and fall experience about the same amount of rainfall. Clouds again are the rule, with either rain or showers occurring every two or three days. Wet spells of several days duration are a distinct possibility. The cool moisture provides the ideal condition for the giant coast redwoods which are found here in great quantities.

SUMMER

Summers remain cool as breezes blow in from the Pacific. Afternoon temperatures usually get no higher than the low 60s and at night drop

CALIFORNIA
AREA No.12   CHART No. 34
APRIL 1ST to NOVEMBER 1ST

Forecast: Generally delightful-dry, sunny thru-out mid-season. Some precipitation in early and late transition periods.

Crescent City

OREGON

NEVADA

Redwoods
Redwood Trail
Eureka
Leggett

Mt Shasta 14,162 Ft.

Lake Tahoe

Santa Rosa
Sacramento
Vineyards

San Francisco
Cable cars
Grapes
San Jose
Santa Cruz

Yosemite Nat Park

Kings Canyon
Fresno
Sequoia Nat Park
Vineyards
Death Valley

SAN JOAQUIN VALLEY
A vast area of fruits & vegetables

Bakersfield

Santa Barbara
Lancaster

Citrus

Los Angeles
Long Beach

Palm Springs
Indio
Date Festival

E.D. Powers

San Diego

Imperial Valley
California's
Market basket

Forecast:
Coastal (north) overcast mild & pleasant
Great Valley very dry sunny often 100° hot-hot?
Mts & over 2000 ft. little rain nights cool.

Eureka 19"
Red Bluff 39.5"
41.5"
Sacramento
S.F.
20.5"
Fresno
21"
Bakersfield
18"
9"
L.A.
San Diego

Aver. annual total inches of precipitation.

50   100
Scale of miles

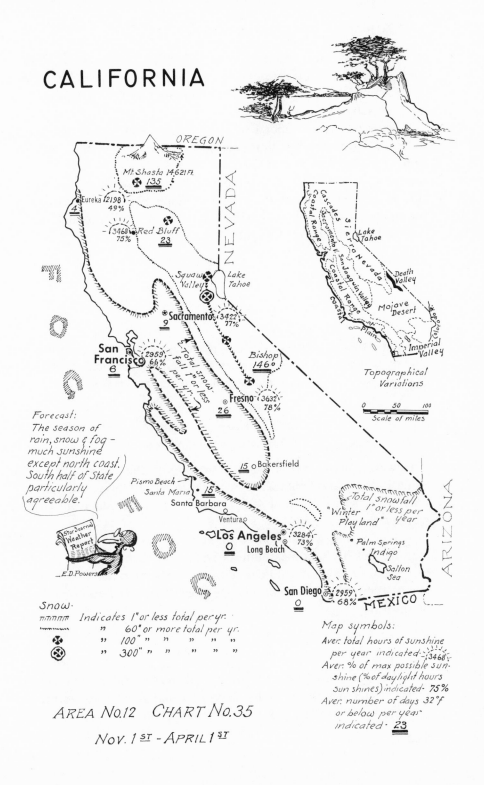

CALIFORNIA

OREGON

Mt. Shasta 14,621 Ft.
⊕ 135

Eureka (2198)
49%
4

(3468) · Red Bluff
75%
23

NEVADA

Squaw
Valley
⊗ · Lake
Tahoe

⊗ Sacramento
9

(3422)
77%

San Francisco (2959)
66%
6

Bishop
146 ○

Fresno
○ (3632)
78%
26

Total snow fall 1" or less per yr.

15 ○ Bakersfield

Forecast:
The season of
rain, snow & fog –
much sunshine
except north coast.
South half of State
particularly
agreeable!

Pismo Beach ·
Santa Maria · 15 ·
Santa Barbara
Ventura ○

Los Angeles ○ (3284)
73%
Long Beach

○ San Diego
○ (2959)
68%

Cascades
Coastal Range
Sacramento V.
San Joaquin Valley
Sierra Nevada
Lake Tahoe
Death Valley
Mojave Desert
Coastal Range
Coastal
Plain
Imperial Valley

Topographical
Variations

0   50   100
Scale of miles

Total snowfall
1" or less per
"Winter
Play land"   year

· Palm Springs
· Indigo
Salton
Sea

ARIZONA

MEXICO

Star Journal
Weather
Report
_E.D.Powers.

Snow·
mmmm *Indicates 1" or less total per yr.*
   " 60" or more total per yr.
⊕ " 100" " " " " "
⊗ " 300" " " " " "

Map symbols:
Aver. total hours of sunshine
per year indicated – (3468)
Aver. % of max possible sun-
shine (% of daylight hours
sun shines) indicated – 75%
Aver. number of days 32°f
or below per year·
indicated · 23

AREA No.12   CHART No.35
Nov. 1 ST – APRIL 1 ST

to the low 50s. Heat waves are very rare but upon occasion the thermometer will soar up to the upper 90s and can even top 100°. There is occasional light rain but it is not enough to interfere with most outdoor activities.

Memorial Day to Labor Day is the tourist season despite frequent fog. Folks from the San Francisco area and elsewhere also enjoy this rather unspoiled rugged stretch from the Russian River up through the redwood country. For the combination of most favorable weather and smallest crowds, perhaps May–June or September–October would be the choice periods. Redwood Highway, a major scenic wonder, is at its best when autumn color brightens the countryside.

Midcoast, San Francisco to Pismo Beach

WINTER

As one travels southward along the rugged coast, temperatures get slightly warmer but are still greatly modified by the Pacific Ocean. For example, during January maximum temperatures reached are:

> San Francisco —mid 50s
> Santa Cruz   —about 60°
> Santa Maria   —low 60s

However, nighttime temperatures drop into the chilly upper 30s at Santa Barbara and Santa Cruz, whereas in San Francisco they usually remain in the low to mid 40s. December and February are slightly warmer than January.

Winter is the wet season, with light to moderate rainfalls. They increase steadily to a maximum in December then gradually decrease to an almost rainless summer. During the winter the sun is behind clouds as often as it is visible and one must expect showers every second or third day.

FALL AND SPRING

Fall temperatures are four or five degrees warmer than those in spring. Reference to the tables show that maximum temperatures experienced are in the mid to upper 60s and minimums are in the mid to low 40s with a very slow warming trend through the spring. Deviations from these temperatures are rare.

As spring progresses, the frequency of showers decreases so that by late spring one expects showers only once a week.

During the fall the frequency of showers increases from almost none in September to about one to two rainy days per week in November. Thunderstorms are practically non-existent along the coast; however, rains in the late fall may be associated with severe Pacific storms moving inland. As a result, moderate to heavy rains may be experienced along with strong gusty winds. The much photographed trees at Monterey, bent one-sided by the ocean winds, are ample evidence of the force and persistency of these blasts.

SUMMER

This is the season when the coast is at its best. It's a glorious sight to drive along Highway Number 1, following the shoreline, on a sunny day in early summer. The beautiful sparkling Pacific lies on one side and a carpet of flowers on the other, and the entire coast seems to be air conditioned. Maximum temperatures remain in the very comfortable low to mid 70s and drop to the upper 40s and low 50s at night. Heat waves, although uncommon, do occur lifting afternoon temperatures to near 90°. They are usually of short duration and the pleasant summer temperatures soon return.

Swimming in the Pacific Ocean in this area can be a shivery experience, due to the very low temperature of the water, and rubber suits are almost a must. This condition is caused by a phenomenon called "upwelling," whereby water from the cold ocean depths is brought to the surface. This is why this area has such cool temperatures, especially noticeable during the summer months.

Rainfall is almost completely absent throughout the entire summer. There may be a few sprinkles but the amounts are too insignificant to measure. While many think of this as a land of winter sunshine, it is fast becoming recognized as an almost ideal year around place to live. Lovely Santa Barbara has long enjoyed that reputation.

Coastal Region—Santa Barbara to San Diego

WINTER

Winters are mild and pleasant with abundant sunshine. January temperatures are in the mid 60s during the afternoon and the low 40s at night. December and February are about two or three degrees warmer.

Winter is the rainy season along the southern coast of California. However, the amount of rain is not substantial; showers are moderate and bunched into periods of a few days separated by many days of brilliant sunshine. Two such spells per month would be about average.

SPRING AND FALL

Spring and fall weather is pleasant. The afternoon temperatures in April are in the low 70s and upper 60s and drop into the low 50s and upper 40s at night. Day to day variations are small. March is about two degrees cooler than April, May is about two degrees warmer than April.

Early fall temperatures are in the very pleasant low 80s with about 60° at night, but gradually cool to the high 70s in October and low 70s in November. Nights also get progressively cooler.

March (the rainiest month), experiences about two rainy periods of a few days each, but the rest of the month is very agreeable. Rain decreases during April and May is almost perfect, with showers on one or two days in the entire month.

The fall season is also very pleasant. There may be one day of showers in September and perhaps two in October; November can have one rainy period of a few days only. Thunderstorms are practically nonexistent.

About midway between Los Angeles and San Diego is the lovely Mission of San Juan Capistrano where the legendary swallows of Capistrano arrive each year on March 19 and fly south on October 23.

SUMMER

Summers are pleasantly dry, warm and sunny with low humidities. Temperatures are comfortable, ranging from the low to mid 80s in the afternoon to the low 60s to upper 50s at night. Humidity in the San Diego area is slightly higher at night. San Diego is a popular retirement and vacation spot. Canadians and northerners migrate here for the winter months, and refugees from the desert lands around Phoenix show up for the summer. Threre is practically no rain in summer and skies are clear almost 80% of the time. About 90% of the world's crop of poinsettias is grown in this area; these flowers are glorious at top color in December.

Smog: This entire area from Santa Barbara experiences fog which produces smog when combined with air pollutants. The intensity varies with the quantity of pollutants, being very much more serious in the Los Angeles environs than any other section. San Diego, which has little industry, suffers smog only very occasionally.

Above the 2000 foot level of the Sierra Nevadas and the Coast Ranges

WINTER

Skiers will remember that mountain temperatures vary with altitudes and will drop about three degrees for every 1000 feet of elevation. Afternoon temperatures in the northeastern areas of Mount Shasta and Yreka are usually in the low to mid 40s and the mid 20s at night during December and January. However, the cold blasts can barrel in from the north dropping the temperatures to single numbers. A warming trend begins in February with day figures near 50° and chilly 20s at night. In the Redding area, the readings will be 10 to 20° higher than the places mentioned above. Lake Tahoe has rather cold winter temperatures, seldom higher than the upper 30s in the day and the mid teens at night through January and February and only a few degrees higher in December. The occasional cold wave can generate readings well below zero, but the discomfort is greatly reduced because the air is dry.

Further south, near Bishop, afternoon readings in the mid 50s are common through December, January, and February. Night temperatures are in the mid 20s. Zero and slightly below temperatures have been recorded in cold spells.

Winter is the season of precipitation in California. Along the coast, it's too warm for much snow and most falls as rain. Inland, however,

215

through the Sierra Nevadas and Cascade ranges, there is heavy snowfall of at least 16 feet per year and in many parts much more. Over large areas of these mountains the mean exceeds 40 feet per year. It is impossible to pinpoint actual snowfall amounts but heaviest snowfalls occur high in the mountains and generally decrease as you drop in elevation.

Spring and Fall

Temperatures during fall are some 6 to 8° warmer than corresponding ones in the spring. Throughout March expect readings in the upper 50s during afternoons in northern California, and even into the low to mid 60s as you travel southward. Higher elevations, such as Mount Shasta, have cooler temperatures with highs around 50°. At night temperatures drop to near 30° in the north and into the mid 40s in the Redding area further south.

April and May see afternoon temperatures about eight degrees each month above the March highs. Because of the dryness of the air, night temperatures drop into the 30s in the north and into the upper 40s and low 50s around Redding.

Rain and snow occur mainly during early and middle spring, tapering off into summer and late fall and increasing toward winter. During March precipitation usually falls as snow, especially in the higher elevations. In the southern Sierras, however, rainfall becomes the dominant form especially in late March. Snow is still possible during April, but rain showers generally occur and by May all precipitation is in the form of rain showers. Snow or rain can be expected about every third day during early spring, decreasing steadily as late spring approaches. Melting snows, however, often delay the start of spring activities in the mountains and some roads are not opened until late May.

Early fall assures excellent weather. Rainfall is scarse and temperatures comfortable. In October the possibility of rain showers or snow increases. Many roads are closed by the end of October, and November experiences rather frequent periods of precipitation with snow a definite threat, especially in the higher elevations.

SUMMER

Again altitude is a factor. Lower elevations in the Sierras generally experience temperatures in the low 90s with some places, such as Redding, reaching the upper 90s. Higher elevations have temperatures remaining in the low 80s during the afternoon. Since the air is so dry, temperatures in the 90s do not feel too uncomfortable. Pleasant sleeping weather can also be expected as night temperatures drop into the upper 40s and low 50s in most areas. Warm clothing is necessary as evenings cool off rather rapidly, and early mornings are also extremely cool.

Summer weather conditions are usually ideal for vacationing. Rain is almost non-existent, and when it occurs it is usually in the form of brief afternoon thunderstorms, particularly in the higher mountains. Showers of this nature may be expected only once every two weeks and pose no threat of ruining a vacation.

The Great Central Valley—Red Bluff, Sacramento, Merced, Fresno, Hanford, Chico, and Bakersfield

WINTER

Afternoon temperatures usually reach the mid 50s in the valley and dip into the mid to upper 30s at night during December and January. Variations in these temperatures are usually quite small; however, temperatures can drop into the 20s on occasion. February shows temperatures about four degrees higher.

Rainfall in the valley decreases as one travels southward. In the northern areas (Sacramento, Red Bluff) rain is quite abundant, occurring as brisk showers about every third day during December, January, and February. Snowfall is rare and only light flurries. Further south (Bakersfield) rainfall is far less frequent with showers about once a week, and traces of snow are rare. Clouds are common, especially through December and January, with more sunshine possible during February.

Spring and Fall

Fall temperatures are five or six degrees warmer than corresponding ones in spring. March readings will be in the mid to upper 60s in the afternoon and the low to mid 40s at night. April figures will be in the low to mid 70s and upper to low 40s at night. May is much warmer with temperatures of low to mid 80s and pleasant mid 50s at night.

Winter gets the most rainfall, both in amount and frequency. Showers may occur twice a week during early spring and decrease to about once every two weeks by the end of May. There is almost no rain in early fall, but by November there is about one shower a week and showers become heavier as winter approaches.

Summer

In a matter of less than a hundred miles, one passes from the cool temperatures along the coast to the very hot temperatures in the Central Valley. This is because the high Coast Range bars the cool foggy weather from the valley beyond. However, a strong sea breeze finds its way into part of the valley through a gap in the mountains. This cool air current spreads north up the Sacramento Valley and south down the San Joaquin Valley. Its influence on the temperature is very evident. Stockton, situated opposite this opening to the sea, enjoys pleasant weather but temperatures increase markedly both northward and southward in the valleys from that point.

The rest of the Central Valley is famous for its hot, dry weather. Temperatures on most summer afternoons reach the upper 90s to 100° mark. Temperatures over 110° occasionally occur. Even though the air is extremely dry, a great deal of discomfort is experienced during the day, but temperatures cool off rapidly once the sun sets, reaching the comfortable 60s at night. Nighttime hours may be more comfortable for traveling in this section of California than during the heat of day. Most locations in the Valley experience about the same high temperature ranges.

Summer precipitation is very light and can hardly be measured. Conversely, there is a great abundance of sunshine.

The Mojave–Colorado Desert

WINTER

Daytime temperatures near 60° are common during December, January and February dropping into the thirties at night.

In recent years this golden area has become a world renowned funland. With Palm Springs as the hub, there are a number of resort sun spots for golf, loafing and sunning. Indio, the date center, produces a greater variety and finer quality of that fruit than does Egypt. It's a main feature of the elaborate spring festival.

Precipitation, which generally occurs during the winter months, is in the form of light rain showers, which quickly evaporate because they're not sufficient to form even a brooklet. No part of the desert receives more than five inches of rainfall per year, most areas getting much less.

SPRING AND FALL

Temperatures through spring and fall are very similar. During the afternoon expect to find the mercury near 80° (slightly higher in May and September) near Barstow and in the mid 80s near Needles. Temperatures at night generally drop about 30° below the afternoon maximum. Precipitation is insignificant.

Table 95	AVERAGE % OF MAXIMUM POSSIBLE SUNSHINE (% of daylight hours)											
Area Number 12 California	Jan	Feb	March	April	May	June	July	Aug	Sept	Oct	Nov	Dec	Year
California													
Eureka	40	44	50	53	54	56	51	46	52	48	42	39	49
Fresno	46	63	72	83	89	94	97	97	93	87	73	47	78
Los Angeles	70	69	70	67	68	69	80	81	80	76	79	72	73
Red Bluff	50	60	65	75	79	86	95	94	89	77	64	50	75
Sacramento	44	57	67	76	82	90	96	95	92	82	65	44	77
San Diego	68	67	68	66	60	60	67	70	70	70	76	71	68
San Francisco	53	57	63	69	70	75	68	63	70	70	62	54	66

SACRAMENTO, CALIFORNIA Elevation 23 Feet Table 96

Month	Average Max.	Average Min.	Extreme Max.	Extreme Min.	THI*	Wind Chill Factor*	Precip. Total	Precip. Snow	Not even 0.01" precip.	More than ½" of snow	Clear	Cloudy	Thunder-storms	Fog	90° or higher	32° or lower	% of possible sunshine	R.H. A.M.	R.H. P.M.	Wind M.P.H.	Wind Direction	Storm Intensity*
Jan	53	37	67	23		26	3	T	20	0	12	19	0	10	0	9	45	88	72	8	SE	R-2
Feb	59	40	76	28			3	T	18	0	15	13	0	6	0	1	60	80	62	8	SSE	R-2
March	65	42	86	28			2	0	23	0	18	13	1	2	0	1	70	70	53	9	SW	R-2
April	71	45	91	34			1	0	24	0	22	8	1	0	0	0	79	62	46	9	SW	R-1
May	78	50	101	37	71		1	0	28	0	25	6	1	0	4	0	85	54	39	10	SW	R-1
June	87	54	115	43	75		T	0	29	0	27	3	0	0	12	0	92	50	34	10	SW	R-1
July	93	57	113	50	79		—	0	31	0	30	1	0	0	24	0	97	48	28	9	SSW	R-1
Aug	92	56	107	49	78		—	0	31	0	30	1	0	0	24	0	96	50	28	9	SW	R-1
Sept	88	55	104	43	75		T	0	29	0	27	3	1	0	13	0	93	52	32	8	SW	R-1
Oct	78	49	99	38	69		1	0	28	0	26	5	0	2	2	0	86	58	40	7	SW	R-1
Nov	64	42	87	26			1	0	24	0	18	12	0	6	0	1	65	78	61	7	NNW	R-2
Dec	55	38	72	24		33	3	T	22	0	14	17	0	9	0	7	46	87	74	7	SSE	R-2
Year	74	47	115	23			16	T	306	0	264	101	4	35	80	20	78	65	47	9	SW	

Notes:
T Indicates "trace"
* For full explanation of (T-H-I) "Temperature Humidity Index," "Wind Chill Factor" and "Storm Intensity," see beginning of Chapter 2.

Average date of first freeze .... December 11
    "   "   " last   "   .... January 24
    "   freeze-free period .... 321 days
10 inches of snow equal approximately one inch of rain.

220

SAN FRANCISCO, CALIFORNIA   Elevation 52 Feet   Table 97

Month	Average Max.	Average Min.	Extreme Max.	Extreme Min.	THI*	Wind Chill Factor*	Precip. Total	Precip. Snow	Not even 0.01" precip.	More than ½" of snow	Clear	Cloudy	Thunder-storms	Fog	90° or higher	32° or lower	% of possible sunshine	Rel. Hum. A.M.	Rel. Hum. P.M.	Wind M.P.H.	Wind Direction	Storm Intensity*
Jan	56	46	79	30			5	T	20	0	20	11	0	4	0	0	56	82	68	7	N	R-3
Feb	57	47	75	36			4	T	17	0	18	10	0	3	0	0	62	80	66	8	W	R-3
March	61	49	83	38			3	T	21	0	22	9	0	2	0	0	69	75	62	9	W	R-2
April	62	50	86	40			1	0	24	0	18	12	0	2	0	0	72	73	64	10	W	R-2
May	63	51	91	44			1	0	27	0	21	10	0	1	0	0	72	75	66	10	W	R-2
June	65	53	101	47			1	0	28	0	22	8	0	1	0	0	73	77	68	11	W	R-1
July	64	53	92	47			—	0	30	0	26	5	0	1	0	0	66	84	75	11	W	R-1
Aug	65	54	96	48			—	0	30	0	27	4	0	1	0	0	66	85	74	11	W	R-1
Sept	69	55	97	48			T	0	29	0	26	4	0	4	1	0	72	77	67	9	W	R-1
Oct	68	54	94	45			1	0	27	0	24	7	0	3	0	0	71	76	62	8	W	R-1
Nov	64	51	86	41			2	0	23	0	24	6	0	4	0	0	64	78	62	6	W	R-2
Dec	58	47	76	35			4	T	21	0	21	10	0	4	0	0	54	81	70	7	N	R-3
Year	63	51	101	30			21	T	298	0	169	96	0	30	1	0	67	79	67	9	W	

Notes:

T Indicates "trace"

* For full explanation of (T-H-I) "Temperature Humidity Index," "Wind Chill Factor" and "Storm Intensity," see beginning of Chapter 2.

Average date of first freeze .... Not Computed
 " " " last " .... " "
 " freeze-free period .... " "
10 inches of snow equal approximately one inch of rain.

BAKERSFIELD, CALIFORNIA — Elevation 494 Feet — Table 98

Month	Avg Max	Avg Min	Extreme Max	Extreme Min	T H I *	Wind Chill Factor*	Precip Total	Precip Snow	Not even 0.01" precip.	More than ½" of snow	Clear	Cloudy	Thunder-storms	Fog	90° or higher	32° or lower	% of possible sunshine	R.H. A.M.	R.H. P.M.	Wind M.P.H.	Wind Direction	Storm Intensity*
Jan	57	37	75	25		35	1	T	25	0	17	14	0	8	0	5	50	74	60	5	NW	R-1
Feb	63	41	80	28			1	T	21	0	17	11	0	3	0	1	68	65	49	6	ENE	R-1
March	69	45	92	31			1	T	24	0	20	11	0	0	0	0	75	52	39	6	NW	R-1
April	76	50	98	38	68		1	0	25	0	22	8	1	0	3	0	83	46	34	7	NW	R-1
May	85	56	102	41	74		T	0	29	0	26	5	0	0	10	0	88	37	24	8	NW	R-1
June	92	62	111	48	79		T	0	30	0	28	2	0	0	17	0	90	35	23	8	NW	—
July	100	68	110	59	87		—	0	31	0	30	1	0	0	29	0	90	31	19	7	WNW	—
Aug	98	66	111	57	84		—	0	31	0	30	1	0	0	27	0	90	33	21	7	NW	—
Sept	92	61	108	50	79		—	0	29	0	29	1	0	0	16	0	90	40	27	6	WNW	R-1
Oct	81	53	97	45	74		T	0	29	0	27	4	0	0	4	0	80	43	31	5	NW	R-1
Nov	69	43	90	30			T	0	27	0	22	8	0	3	0	1	70	65	51	5	ENE	R-1
Dec	59	39	75	25		36	1	T	25	0	15	16	0	8	0	5	60	74	62	5	ENE	R-1
Year	78	52	111	25			6	T	327	0	283	82	1	22	106	12	80	49	37	6	NW	1

Average date of first freeze .... November 28
   " " " last .... February 14
   " freeze-free period .... 287 days
10 inches of snow equal approximately one inch of rain.

Notes:

T Indicates "trace"

* For full explanation of (T-H-I) "Temperature Humidity Index;" "Wind Chill Factor" and "Storm Intensity," see beginning of Chapter 2.

LOS ANGELES, CALIFORNIA

Elevation 312 Feet — Table 99

Month	Temperatures — Average Max.	Average Min.	Extreme Max.	Extreme Min.	THI*	Wind Chill Factor*	Precipitation in inches — Total	Snow	Avg. days — Not even 0.01" precip.	More than ½" of snow	Clear	Cloudy	Thunder-storms	Fog	90° or higher	32° or lower	% of possible sunshine	Rel. Humidity A.M.	P.M.	Wind M.P.H.	Direction	Storm Intensity*
Jan	65	47	86	28			3	T	25	0	22	9	1	2	0	0	72	51	50	7	NE	R-2
Feb	66	48	90	34			3	T	23	0	19	9	1	2	0	0	72	54	52	7	W	R-2
March	69	50	90	38			2	0	25	0	22	9	1	1	0	0	73	52	52	7	W	R-2
April	71	53	99	41	69		1	0	26	0	20	10	1	1	1	0	68	53	54	7	W	R-2
May	74	56	102	46	69		T	0	29	0	24	7	0	1	1	0	66	56	55	6	W	R-1
June	77	59	104	50	72		T	0	29	0	25	5	0	1	1	0	65	59	56	6	W	R-1
July	83	63	103	54	77		—	0	31	0	30	1	0	1	3	0	82	54	53	5	W	R-1
Aug	83	63	103	53	77		—	0	30	0	29	2	0	1	4	0	82	56	55	5	W	R-1
Sept	83	61	110	51	76		T	0	29	0	28	2	0	1	6	0	78	52	54	5	W	R-1
Oct	77	57	104	46	72		T	0	29	0	25	6	0	3	3	0	74	55	56	6	W	R-2
Nov	73	52	100	39			1	0	26	0	24	6	1	2	1	0	74	45	49	6	W	R-2
Dec	68	49	89	32			3	T	26	0	23	8	1	2	0	0	71	45	50	7	NE	R-2
Year	74	55	110	28			15	T	328	0	291	74	6	17	18	0	73	53	53	6	W	

Notes:

T Indicates "trace"

* For full explanation of (T-H-I) "Temperature Humidity Index," "Wind Chill Factor" and "Storm Intensity," see beginning of Chapter 2.

Average date of first freeze ....Not Computed

" " last " .... " "

" freeze-free period .... " "

10 inches of snow equal approximately one inch of rain.

223

SAN DIEGO, CALIFORNIA — Elevation 19 Feet — Table 100

Month	Temperatures Average Max.	Average Min.	Extreme Max.	Extreme Min.	THI*	Wind Chill Factor*	Precip. Total	Snow	Not even 0.01" precip.	More than ½" of snow	Clear	Cloudy	Thunder-storms	Fog	90° or higher	32° or lower	% of possible sunshine	R.H. A.M.	R.H. P.M.	Wind M.P.H. Direction	Storm Intensity*
Jan	65	45	86	31			2	T	24	0	20	11	0	4	0	0	70	53	55	6 NE	R-2
Feb	65	47	85	38			2	0	21	0	18	10	0	3	0	0	72	56	57	6 WNW	R-2
March	68	50	85	42			2	0	24	0	22	9	0	2	0	0	71	58	58	7 WNW	R-2
April	69	54	91	44			1	0	25	0	19	11	0	2	0	0	63	59	60	8 WNW	R-2
May	71	57	91	48			1	0	29	0	21	10	0	1	0	0	59	65	64	8 WNW	R-1
June	73	60	82	51			T	0	29	0	23	7	0	1	0	0	55	70	67	7 SSW	R-1
July	77	63	88	57	72		T	0	31	0	27	4	0	1	0	0	68	70	67	7 WNW	—
Aug	78	66	89	60	73		T	0	31	0	27	4	0	1	0	0	69	67	66	7 WNW	—
Sept	78	62	111	56	72		T	0	29	0	26	4	0	3	1	0	69	65	65	7 NW	R-1
Oct	74	58	107	48			1	0	28	0	24	7	0	4	1	0	67	59	62	6 WNW	R-2
Nov	72	51	97	38			1	0	26	0	24	6	0	4	0	0	74	59	64	6 NE	R-2
Dec	67	47	88	36			2	T	25	0	22	9	0	5	0	0	71	56	59	5 NE	R-2
Year	71	55					10	T	322	0	273	92	0	29	3	0	67	61	62	7 WNW	

Notes:

T Indicates "trace"

* For full explanation of (T-H-I) "Temperature Humidity Index," "Wind Chill Factor" and "Storm Intensity," see beginning of Chapter 2.

Average date of first freeze . . . . Not Computed
     "   "   " last   "   . . . .   "   "   "
     "   " freeze-free period   . . . .   "   "   "
10 inches of snow equal approximately one inch of rain.

SUMMER

The desert is extremely hot and dry. Afternoon temperatures over 100° are the rule. Highest recorded shade temperature is 134° at Greenland Ranch, Death Valley, which is 282 feet below sea level. It's interesting that the lowest spot in this country is only 84 miles from the highest point in the contiguous United States—Mount Whitney at 14,495 feet.

Although Death Valley is a place to avoid in summer, midday temperatures from early November through the end of March are usually a pleasant 65 to 75 degrees and 40 to 50 degrees at night. There will be only 63 or 74 cloudy days in winter and perhaps one in ten are overcast. Two inches of rainfall a year can't stir up much cloudy weather!

At night expect temperatures to drop into the comfortable 60s near Barstow but remain in the upper 70s to near 80° in the Needles area. Occasionally, temperatures remain in the 90s at night.

Rainfall is almost non-existent in the desert areas. In the surrounding mountains, however, heavy rains of cloudburst proportions wash loads of materials into the valley, forming huge, alluvial fans on both sides of the valley. Sunshine is abundant.

3

Alaska

See Chart No. 36

Did you know that the Fairbanks area averages more spring sunshine than New York City, and gets 22 hours of daylight on June 21st? It's not unusual at almost any time of the year to read that temperatures in Alaskan cities are higher than in Chicago or Boston.

But Alaska will probably not replace Florida as a winter resort, even though increasing numbers of visitors arrive in Alaska every summer, undaunted by the great distance, the permafrost and the tales of Jack London. A substantial number of young couples also become new residents each year, but it is not a very pleasant place for older retirees. The median age of the entire population is only 18.5 and less than 5% are over 65 years old.

The climate of Alaska can be divided into zones determined by such factors as lofty mountains, prevailing winds, warm ocean currents and frozen seas. The U.S. Weather Bureau treats it in eight such sections.

The *Southeast* or *Panhandle* coast consists of a multi-chain of islands backed by a narrow ribbon of mainland that borders Canada's British Columbia and a corner of the Yukon Territory. This is an area of high precipitation, mostly falling as rain, but with no very great temperature extremes. It has been likened to the Puget Sound region and with good reason. Both coasts extend north and south and are washed by the warm waters of tropical born ocean streams. In southeast Alaska, the prevailing southwesterly winds, coming in moist from the warm Pacific, drop precipitation in varying amounts from 30 to 220 inches depending upon local topigraphical conditions. Both of these

226

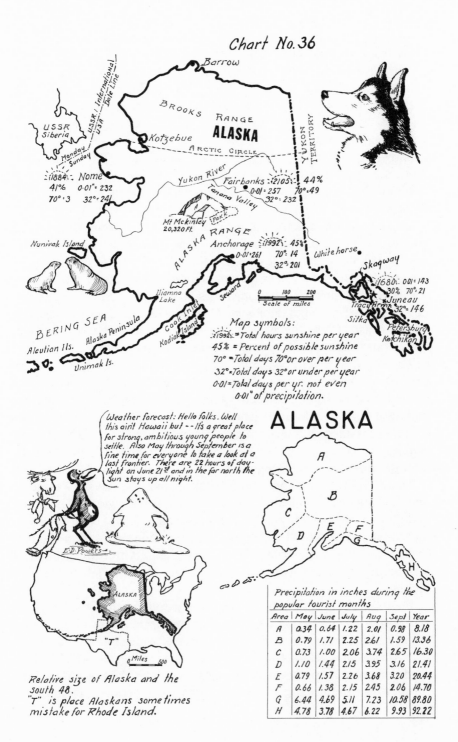

Chart No. 36

ALASKA

Barrow

USSR Siberia

USSR International Date Line
USA

BROOKS RANGE

Kotzebue

Monday Sunday

ARCTIC CIRCLE

YUKON TERRITORY

Nome
1884
41% 0·01"=232
70°=3 32°=241

Yukon River

Fairbanks 2105 44%
0·01=257 70°=49
32°=232

Tanana Valley

Mt McKinley
20,320 Ft. Park

ALASKA RANGE

Anchorage 1992 45%
0·01=261 70°=14
32°=201

Whitehorse

Skagway

1680 0·01=143
30% 70°=21
Juneau
32°=146

Nunivak Island

Iliamna Lake

Seward

Tracy Arm
Sitka Petersburg
Ketchikan

BERING SEA

Cook Inlet
Kodiak Island

Aleutian Ils.

Alaska Peninsula

Unimak Is.

0 100 200
Scale of miles

Map symbols:

1992 = Total hours sunshine per year
45% = Percent of possible sunshine
70° = Total days 70° or over per year
32° = Total days 32° or under per year
0·01 = Total days per yr. not even
0·01" of precipitation.

(Weather forecast: Hello folks. Well this ain't Hawaii but -- It's a great place for strong, ambitious young people to settle. Also May through September is a fine time for everyone to take a look at a last frontier. There are 22 hours of daylight on June 21st and in the far north the Sun stays up all night.

E.D. Powers

ALASKA

ALASKA

A
B
C
D E F G
H

"T"

0 Miles 500

Relative size of Alaska and the
south 48.
"T" is place Alaskans sometimes
mistake for Rhode Island.

Precipitation in inches during the popular tourist months					
Area	May	June	July	Aug	Sept	Year
A	0.34	0.64	1.22	2.01	0.98	8.18
B	0.79	1.71	2.25	2.61	1.59	13.36
C	0.73	1.00	2.06	3.74	2.65	16.30
D	1.10	1.44	2.15	3.95	3.16	21.41
E	0.79	1.57	2.26	3.68	3.20	20.44
F	0.66	1.38	2.15	2.45	2.06	14.70
G	6.44	4.69	5.11	7.23	10.58	89.80
H	4.78	3.78	4.67	6.22	9.93	92.22

coastal areas are backed by high ranges of mountains which protect them from the cold northerly winds.

During some winters, severe northern gales reach the Alaskan Panhandle coast through gaps and river valleys. These frigid land winds, with velocities up to 125 m.p.h. can do great damage in the local areas into which they are funneled, though places only a mile or two away may be untouched.

Snowfall can vary from 26 inches to as much as 290 inches per year. Snowfalls generally occur in the intervals between changes of wind direction, but are soon followed by rain which quickly washes away the snow.

There is cloudy weather but most of the summer, from May through August, is sunny and pleasant. The warm Alaskan current which follows north along the coast, tempers the air and modifies the temperature range. This benevolent stream is an off-shoot of the much larger warm North Pacific Drift, sometimes called the Japan Current, that flows eastward through the Pacific Ocean headed directly for the Puget Sound coast. There the Alaska Current branches off from the main stream and heads northward along the Canadian shore.

The Alaska Current continues westerly along the south shore of the Gulf of Alaska until it loses its heat and is diluted with the cold waters around Kodiak Island. This coastline is rather regular with few inlets and little signs of habitation from the Panhandle almost to Cook Inlet. The shore is lined with sharp ragged ice cliffs discolored with earth and is quite unattractive. The weather rather resembles that of the Panhandle but with a bit less precipitation.

The *Southwest* area includes Kodiak Island, the Alaska Peninsula, and the long chain of bleak Aleutian Islands. It is a region of few people and almost perpetual fogs. The raw damp air, coupled with the winds and sea storms, makes it forbidding. Precipitation varies from about 30 inches annually in the Pribilof Islands, the famous breeding grounds of seal herds, to 60 inches at Unalaska where the snowfall is about 80 inches.

West Central Alaska is the broad flat portion, north of the Peninsula and bordering the Bering Sea. This is a sportsman's paradise. It's the home and nesting ground of many species of waterfowl including that true Alaskan native, the Emperor goose. The melting snow forms thousands of pools as permafrost prevents the water from soaking down through the tundra turf. There are also wide expanses of mosquito infested muskeg swamp. This almost uninhabited rolling country, with low mountains and scattered trees, makes an ideal home site for wildlife.

The Eskimos and Indians live off the land and sea, relatively undisturbed by rapidly changing conditions in more populated areas. There are just two small towns along this coast, and only trading posts in the interior.

Bethel, which was founded by Moravian missionaries, is a headquarters for sportsmen, but most activity centers around government operations. This section averages about 18 inches of precipitation a year and has a temperature range of 8° to 55°. Dillingham conditions are much the same but here the weather is tempered a little by proximity to the warm ocean currents on the south side of the peninsula. It gets 25 inches total precipitation and the temperatures are from 15° to 55°. Both towns have occasionally experienced 90° and −40° temperature readings. There are frequent strong winds. Dillingham is the biggest red salmon fishing ground in the world, and sportsmen hunt with gun or camera for bear, moose and caribou.

Central Alaska This very large area, with Fairbanks in the center, experiences the widest temperature extremes in all Alaska—from 76° below zero to 100° above. It also registers the lowest readings, being colder than even the far north Arctic region, and experiences the hottest summers. Annual precipitation is a low 14-inch average.

This section includes one of Alaska's three large agricultural areas, the bountiful Tanana Valley which extends down past Big Delta. Snow is plentiful and the winters are very cold.

Arctic Alaska, which includes the Northwest, is the land of the Eskimo and Indian. Precipitation is low, averaging about 18 inches in Nome and along the coast, with only 8 to 9 inches inland and in the northerly portion. Nome temperatures range from a January low average of 4° to a midsummer mean of 50°, with the very occasional extremes of −45° low or an 84 degree high.

Although this is the land of the midnight sun, it is not exactly a six months light and six months dark deal. At Barrow, for instance, the sun isn't visible for 51 days; but for 73 days, from May 16 to July 27, it shines twenty-four hours a day.

Barrow, at the northmost point, is the largest Eskimo settlement in the world. There is only about 4 inches total annual precipitation but with the melting snow, this makes this little frontier village a muddy place in mid-summer. Close by is the Will Rogers-Wiley Post monument.

The very high living costs, common to all Alaska, may be even a bit steeper way up here. Residents suffer from the fact that ships arrive loaded but, lacking exports, return empty; meaning double freight rates for very long hauls.

Alaska extends through four time zones and its almost 34,000 miles of coastline is longer than the total of the other forty-nine states. Although the Panhandle section in the extreme southeast is quite small, only about 100 miles by 400 miles, it is very popular with tourists. Perhaps the main reason for this is the spectacular Inside Passage which winds its way among the thousands of islands and high cliffs. Its misty but quite mild climate is quite similar to that of Great Britain. Actually Ketchikan is at almost exactly the same latitude as Edinburgh in south Scotland. The almost 1000 mile water trip between Seattle and Skagway can be a delight and an outstanding highlight of a visit to Alaska. Winters are not severe and summer days, while often damp, are pleasant. The precipitation for this whole area averages almost 100 inches per year, being a bit heavier in winter, but there is considerable variation throughout the coastline. Ketchikan, in the lower section, is deluged with 150 inches, but Haines at the far north end gets only 60 inches per year.

Most of the main towns in the Panhandle are on islands which makes the boat trip a most convenient way to see them. When fog doesn't blot out visibility, a small cruise ship may go into Tracy Arm. This is indeed a scenic treat: a thirty-mile fiord, only a half-mile wide, is cut through sheer rock walls! There are little patches of flower covered meadow and mountain goats on the high cliffs but no sign of human habitation. At the head of this magnificent gorge are the two glistening white glaciers, the North and South Sawyers. Massive ice blocks float like sparkling jewels down the fiord. Dense fog often prevents ships entering the gorge but when the sun shines, Tracy Arm is a sight that will long be remembered.

In addition to magnificent scenery, there are a number of interesting communities along the waterway. Sitka on Baranof Island, once known as the Paris of the Pacific, was the gay Russian capital. Although it retains a little old world flavor, it more resembles a small New England coastal village. It is a favorite fishing center and attracts hunters for deer, bear and mountain goat. Being one of the few places facing directly onto the Pacific, Sitka is not a stop on all cruises as the boats generally stay in the protected Inside Passage.

Petersburg, a little fishing port on Mitkof Island, was settled by Scandinavians and still celebrates Norwegian Independence Day on May 17th. This most attractive little village depends almost entirely on fishing, mostly salmon and shrimp. It is surrounded by the majestic Tongass National Park.

Ketchikan is the salmon center of the world and should not be missed. Like Juneau it has a Salmon Derby that attracts many enthusiasts. Its 150 inch annual precipitation (mostly rain) and overcast skies make it an unlikely stop-off spot for more than a few days.

Juneau, clinging at the foot of lofty mountains about half way up the archipelago, is the capital. It boasts a famous Salmon Derby and a fascinating museum of Eskimo and Indian crafts. Juneau is almost isolated, being hemmed in by high mountains with only a few short roads that really go nowhere. There are many glaciers in Alaska but perhaps the one most photographed is the Mendenhall, a short distance from Juneau. It enjoys a mild climate and the 83 inches of rain and 115 inches of snow, which soon disappear, does not frighten off either tourist or sportsman. It's interesting to note that total precipitation is 35 inches less at Juneau's airport, which is only nine miles from the city center.

About 100 miles north is the little town of Haines, where many leave the cruise. It enjoys a very pleasant summer climate with the thermometer often at a comfortable 85°. The 60 inch annual precipitation is low for the area. This is the only town in the southeast with a connecting road to the Alaska Highway. The Strawberry Festival, held in late June or early July, is a treat.

Skagway, a short distance up at the head of the Lynn Canal, is at the very upper end of this district. It has much the same weather as Haines except a little less rain, and usually less than 2″ each of the months, from March through August. A worthwhile jaunt is the run to Whitehorse and the Canadian interior, on the picturesque narrow gauge White Pass and Yukon Railway. The pass was once an outlet to the famous Klondike gold country.

If there are no school age children in the party, it's worth considering the off-season period for a visit to Alaska. Some of the cruise ships offer a 25% discount from late May through early or mid-June and from late August through early September. Apart from the savings, there is greater ease in arranging accommodations and general transportation. In the earlier period, the days are getting long and spring flowers cover many fields. There is also far less likelihood of being annoyed by mosquitoes and other insects, which are present in great hordes later in the year. The autumn period is often sunnier and drier than mid-season. In addition, the foliage colors in September rival New England. Check with your travel agent and for additional information write to the Alaska Travel Division, Pouch E, Juneau, Alaska, 99801.

Most visitors fly from the Skagway or Juneau areas directly to Anchorage, which is the financial and commercial center of Alaska. It is one of the fastest growing cities in the United States on a percentage basis.

Although facing on Cook Inlet and close to the wet coastline, Anchorage averages only 14.5 inches of total precipitation per year, 60 inches as snow. (Ten inches of snow equal about one inch of rain.) It is most favorably situated, being protected about three quarters of the way around by an almost continuous wall of high mountains. They shield it from the moist southerly sea breezes and the cold blasts from the north, so that it is drier than the nearby coast and far milder than the Fairbanks area to the north. This is another example of how topography can radically effect places only a short distance apart.

The January temperatures usually range from about 10 to 28° and summer temperatures average out in the 60s. The growing season can be 100 to 150 days, which explains the giant sized berries and vegetables and makes the abundant Matanuska Valley one of the three principal agricultural centers of Alaska.

The most interesting way to see some of the inland country is the railway trip from Anchorage to Fairbanks with a stop at Mt. McKinley Park. Fairbanks is the farthest north large town in America. While much of the housing and buildings are quite new, the town has retained more of the old pioneer atmosphere than Anchorage. It is an excellent headquarters for visitors and is well worth a several day stopover.

Fairbanks takes you into a different climate zone—wide temperature ranges with sudden changes, hot summers and severe winters. The Alaska Range bars the south breezes from the wet south coast. The total annual precipitation is only 12 inches, but 90 inches falls as snow. There is a much wider temperature swing than experienced either in the south or the far north. The July average is about 60° and that in January is −11°. The most extreme temperatures recorded were 99°

JUNEAU, ALASKA — Elevation 17 Feet — Table 101

Month	Average Max.	Average Min.	Extreme Max.	Extreme Min.	THI*	Wind Chill Factor*	Precip. Total	Precip. Snow	Not even 0.01" precip.	More than ½" of snow	Clear	Cloudy	Thunder-storms	Fog	900° or higher	32° or lower	% of possible sunshine	R.H. A.M.	R.H. P.M.	Wind M.P.H.	Wind Direction	Storm Intensity*
Jan			57	−20		10	4	22	13	6	7	24	0	2	0	26	30	80	78	9	ESE	S, R-3
Feb			49	−12		11	3	22	11	5	7	21	0	2	0	23	31	82	76	9	ESE	S, R-3
March			55	−11		15	3	19	13	4	8	23	0	2	0	25	38	77	68	9	ESE	S, R-3
April			71	6		21	3	6	13	1	7	23	0	1	0	17	40	73	63	9	ESE	R, S-3
May			82	25		30	3	T	14	0	8	23	0	1	0	5	37	73	62	8	ESE	R-2
June			84	31			3	0	15	0	8	22	0	0	0	0	33	74	63	8	N	R-2
July			84	36			4	0	14	0	7	24	0	0	0	0	30	80	69	8	N	R-3
Aug			83	27			5	0	13	0	8	23	0	1	0	0	30	83	72	8	N	R-3
Sept			72	26			7	T	10	0	6	24	0	3	0	1	25	87	76	8	N	R-3
Oct			61	16		27	8	1	8	0	5	26	0	3	0	8	18	86	80	10	ESE	R-3
Nov			56	−5		17	6	11	9	4	5	25	0	3	0	19	24	85	81	9	ESE	R, S-3
Dec			54	−21		10	4	23	9	6	5	26	0	2	0	24	18	82	81	10	ESE	S, R-3
Year			84	−21			55	103	152	26	81	284	0	21	0	149	31	80	72	9	ESE	

Notes:

T Indicates "trace"

* For full explanation of (T-H-I) "Temperature Humidity Index;" "Wind Chill Factor" and "Storm Intensity," see beginning of Chapter 2.

Average date of first freeze	October 20
" " " last	April 26
" freeze-free period	178 days

10 inches of snow equal approximately one inch of rain.

ANCHORAGE, ALASKA — Elevation 90 Feet — Table 102

Month	Temperatures Average Max.	Average Min.	Extreme Max.	Extreme Min.	THI*	Wind Chill Factor*	Precip. Total	Precip. Snow	Not even 0.01" precip.	More than ½" of snow	Clear	Cloudy	Thunderstorms	Fog	90° or higher	32° or lower	% of possible sunshine	Rel. Hum. A.M.	Rel. Hum. P.M.	Wind M.P.H.	Wind Direction	Storm Intensity*
Jan	20	4	44	−19		2	1	12	24	4	14	17	0	8	0	31	39	74	73	6	NNE	S, R-2
Feb	26	11	41	−21		6	1	13	24	3	12	16	0	5	0	28	42	75	69	6	N	S, R-2
March	33	17	49	−15		10	1	6	25	3	15	16	0	2	0	24	56	66	56	7	N	S, R-2
April	45	29	62	10		19	T	4	25	1	14	16	0	1	0	21	56	66	52	7	N	R, S-2
May	56	39	75	17		30	1	T	26	0	13	18	0	0	0	5	53	60	47	8	S	R-2
June	64	47	75	40			1	0	23	0	14	16	0	0	0	0	46	69	59	8	S	R-2
July	66	50	78	38			2	0	20	0	12	19	1	0	0	0	43	73	62	7	S	R-2
Aug	63	48	71	37			3	0	16	0	13	18	0	1	0	0	37	77	65	7	S	R-2
Sept	56	40	70	29		37	3	2	16	0	12	18	0	2	0	1	38	83	66	6	NNE	R-2
Oct	43	29	58	2		23	2	8	21	2	12	19	0	2	0	23	41	76	63	6	N	R, S-2
Nov	29	16	53	−8		12	1	11	22	4	12	18	0	4	0	26	37	76	71	6	NNE	S, R-2
Dec	21	6	46	−30		3	1	21	24	5	13	18	0	5	0	30	36	73	73	6	NNE	S, R-2
Year	43	28	78	−30			15	76	266	22	156	209	1	29	0	189	45	72	63	7	N	

Notes:
T Indicates "trace"
* For full explanation of (T-H-I) "Temperature Humidity
Index," "Wind Chill Factor" and "Storm Intensity," see
beginning of Chapter 2.

Average date of first freeze . . . . September 13
 "   "   last   "   . . . . May 18
 freeze-free period . . . . 118 days
10 inches of snow equal approximately one inch of rain.

234

NOME, ALASKA — Elevation 13 Feet — Table 103

Month	Temperatures Average Max.	Average Min.	Extreme Max.	Extreme Min.	T H I*	Wind Chill Factor*	Precipitation Total	Snow	Avg days: Not even 0.01" precip.	More than ½" of snow	Clear	Cloudy	Thunder-storms	Fog	90° or higher	32° or lower	% of possible sunshine	Rel. Hum. A.M.	P.M.	Wind M.P.H.	Direction	Storm Intensity*
Jan	12	−3	43	−39		−31	1	10	21	4	16	15	0	3	0	31	39	77	77	12	E	S-1
Feb	13	−2	47	−42		−31	1	6	19	2	14	14	0	2	0	28	45	75	73	12	NE	S-1
March	16	−1	42	−38		−24	1	9	21	4	15	16	0	2	0	31	48	78	75	11	E	S-1
April	28	14	51	−27		−5	1	7	21	2	14	16	0	2	0	30	48	80	77	11	N	S, R-1
May	41	29	70	−11		13	1	2	23	0	14	17	0	4	0	19	44	80	76	11	N	S, R-1
June	52	39	81	25		27	1	T	21	0	13	17	0	5	0	4	39	81	77	10	WSW	R-1
July	55	44	80	32			2	0	18	0	9	22	0	4	0	0	30	86	82	10	WSW	R-1
Aug	54	44	81	30			4	T	14	0	7	24	0	2	0	1	26	87	82	11	SW	R-2
Sept	48	36	63	17		23	3	1	17	0	10	20	0	1	0	10	28	85	76	12	N	R-2
Oct	35	24	59	−3		8	2	5	22	2	12	19	0	0	0	27	34	80	74	12	N	S, R-1
Nov	23	10	44	−39		−14	1	10	17	3	11	19	0	1	0	30	27	80	79	12	N	S-1
Dec	13	−1	40	−41		−19	1	8	21	3	13	18	0	2	0	31	34	76	76	10	E	S-1
Year	33	20	81	−42			18	57	225	20	148	217	0	26	0	240	37	82	80	11	N	

Average date of first freeze . . . . August 22
   "   "   last . . . . June 12
   "   freeze-free period . . . . 74 days
10 inches of snow equal approximately one inch of rain.

Notes:
T Indicates "trace"
* For full explanation of (T-H-I) "Temperature Humidity Index;" "Wind Chill Factor" and "Storm Intensity," see beginning of Chapter 2.

FAIRBANKS, ALASKA Elevation 436 Feet Table 104

Month	Average Max.	Average Min.	Extreme Max.	Extreme Min.	T H I*	Wind Chill Factor*	Total	Snow	Not even 0.01" precip.	More than ¾" of snow	Clear	Cloudy	Thunder-storms	Fog	90° or higher	32° or lower	% of possible sunshine	R.H. A.M.	R.H. P.M.	Wind M.P.H.	Wind Direction	Storm Intensity*
Jan	-1	-21	38	-56		-21	1	10	23	4	15	16	0	4	0	31		70	73	3	N	S-1
Feb	10	-15	37	-49		-17	1	12	21	3	14	14	0	2	0	28		70	69	4	N	S-1
March	24	-6	48	-46		-8	T	8	24	2	16	15	0	1	0	31		76	62	5	N	S-1
April	42	17	60	-21		12	T	4	26	1	17	13	0	0	0	29		72	57	6	N	S-1
May	59	35	81	-1		31	1	1	24	–	15	16	0	0	0	12		60	44	7	N	R, S-1
June	71	46	91	37			1	T	20	–	15	15	2	0	0	0		62	42	7	SW	R-1
July	72	48	88	37			2	0	18	0	13	18	2	1	0	0		69	54	6	SW	R-1
Aug	65	43	85	30			2	T	17	0	10	21	1	2	0	1		76	53	6	N	R-1
Sept	54	33	80	20		32	1	1	20	–	9	21	0	2	0	6		77	50	6	N	R, S-1
Oct	35	17	61	-15		16	1	9	21	3	10	21	0	2	0	29		81	70	5	N	S, R-1
Nov	13	-6	42	-43		-9	1	11	22	3	12	18	0	2	0	30		83	82	4	N	S-1
Dec	2	-18	42	-56		-21	1	11	24	3	14	17	0	4	0	31		76	76	3	N	S-1
Year	37	14					11	66	260	19	160	205	5	19	0	227		73	61	5	N	

Average date of first freeze .... August 28
" " last " .... May 25
" freeze-free period .... 95 days
10 inches of snow equal approximately one inch of rain.

Notes:
T Indicates "trace"
* For full explanation of (T-H-I) "Temperature Humidity Index," "Wind Chill Factor" and "Storm Intensity," see beginning of Chapter 2.

and −65°. Winter is a cold dark period; dawn begins at about 10:00 A.M. on December 21st and sundown is 3½ hours later at 1:30 P.M. But June 21st has almost 22 hours of daylight between dawn and sunset—when they play the Midnight Sun Baseball Game, which begins at 10:30 P.M., and the midnight sun golfers tee off at midnight.

There is much social and cultural activity in Fairbanks. The beautiful university is well worth a visit, as are several museums. And there is Tanana Valley, the "valley of giants"—that is, giant vegetables, and berries. Would you believe a 61-pound cabbage—three pound potatoes, rhubarb four foot high and rye topping seven feet? The growing season is only 90 days, but it's practically all daylight and the plants don't sleep.

The biggest sports event is the North American Sled Dog Championship Derby in March. Another novel sight for most visitors is to see dozens of small private planes, parked wing to wing around take-off lakes. Pontoons are used in summer and skis in winter. Flying is by far the most common form of transportation and is used by almost 80% of the population.

An interesting side trip out of Fairbanks is the triangular tour with a stop at Nome and an overnight stay at the Eskimo village of Kotzebue, just above the Arctic Circle. When the weather is favorable, which is not every day, you may see the tip of Mt. McKinley poked up through the fog or clouds. Total precipitation is an amazingly low 9½ inches. You can shiver in the winter 30° to 40° below zero or enjoy the 50° average during the tourist season when only a warm sweater is required.

Of the 2200 inhabitants, 800 are Eskimos, 200 are whites associated with the hospital or government activities, and the other 1200 are sled dogs. These will gradually be replaced by the cheaper to maintain snowmobiles, but in the meantime a long string of dogs staked outside of a house is still a status symbol. Twelve or fourteen dogs is a Cadillac or Rolls home; four a Volkswagen house.

We strongly urge that anyone contemplating the journey over the Alaskan Highway take along a copy of "Mile Post" guide, which you can get by writing to Mile Post, Box 2175 Anchorage, Alaska. While you may be annoyed by mosquitoes and other hungry insects and find the weather a bit less salubrious than Hawaii, you need not fear pollution, heat exposure, hay fever, poison ivy or oak. All in all, we think you will agree that Seward's Folly was well worth the $7,200,000, even in 1867 dollars.

4

Hawaii

See Chart No. 37

Hawaii, blessed with a beautiful year-round climate and practically no bad period, has no real tourist season. If there is any best time to avoid crowds, perhaps it might be the months between Christmas and school vacation time. From the viewpoint of weather the month of May should be near perfect. Accommodations may be a little easier to come by at that time and in some cases rates are a bit lower.

In spite of its hula skirts and waving palms, you can find skiing and freezing temperatures—but only above the winter snow line on a few mountain tops. Conversely you may have to wait two years or more to experience a 90° day and then only during August, September or October in Hilo or Lihue.

Most of the Hawaiian Islands lie in the path of the rain-bearing northeast trade winds which blow consistently, but usually gently, about nine months or more a year. Most of their moisture is released on the mountain slopes facing the northeast. The leeward sides of the islands are relatively arid with the total annual rainfall averaging 15 inches or even less in some localities. Precipitation in Hawaii is sometimes described as being *scarce* on the leeward side of the islands, *substantial* where facing the trade winds, and *incessant* on many of the mountain tops where it may reach an annual average of 200, 300 or even the astonishing 471 inches—which adds up to 39 feet of water.

The climate of the Hawaiian archipelago is subtropical rather than tropical, since the islands are quite close to the fringe of the tropics. As a result, they can experience occasional winter storms which bring the greatest amounts of rain to lowland areas, especially in the normally

HAWAII

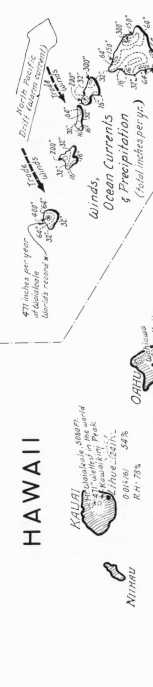

NIIHAU

KAUAI
Waialeale 5080 Ft.
41" Wettest in the world
Kawaikini Peak
Lihue 0·01"=161   54%
R.H. 78%

Winds, Ocean Currents & Precipitation
(Total inches per yr.)

471 inches per year at Waialeale. World's record x   32"-64"

North Pacific Drift (warm current)
Trade Winds

North Equatorial Current (warm)
Trade Winds

OAHU
Waianae Mts   Wahiawa   Kailua
Koolau Mts
Honolulu 0·01"·217   -·304"
R.H. 71%

MOLOKAI

LANAI   Observatory

MAUI
Haleakala Crater
Hana
Puu Alaea, 7662 Ft

KAHOOLAWE

HAWAII
1670   39%
0·01"·78   Hilo   R.H.·84%
Mauna Kea 13,796 Ft
Hualalai
Mauna Loa 13,680 Ft
Kilauea Crater
Kailua   Kona
South Point, 18°56'N
Most southerly point, U.S.A.

Map symbols:-
·304"= Total hrs sunshine per yr.
65%= Percent max possible sunshine
R.H 71%= Relative humidity (aver)
0·01"·217= Total days per yr. not even 0·01 inches of precipitation.

Chart No. 37

Scale of miles
0   10   20   30

Aloha Oe:- Weather forecast - tradewind side of islands moist & lush - lee side dry & sunny. Rain heaviest during winter - little change in temperature throughout year. "A weatherman's life is such an easy one" ♪♫

Diamond Head

SUNSHINE
Total hours per year and percent of maximum possible sunshine

	Jan.	Feb.	Mar.	April	May	June	July	Aug.	Sept.	Oct.	Nov.	Dec.	Year
Hilo	153	135	161	112	106	158	184	134	137	153	106	131	1670
	48%	42%	41%	34%	31%	41%	44%	38%	42%	41%	34%	36%	39%
Honolulu	227	202	250	255	276	280	293	290	257	210	221	211	3041
	62%	64%	60%	62%	64%	66%	67%	70%	70%	68%	63%	60%	65%
Lihue	171	162	176	176	211	246	246	236	246	210	170	161	2411
	48%	48%	48%	46%	51%	60%	58%	59%	67%	58%	51%	49%	54%

E.D.Powers Jr

dry leeward portions. Temperatures vary chiefly with altitude. Thus at the observatory on Maui, just over 10,000 feet above sea level, temperature readings below 20° have been recorded. Down on the sea coast, it is quite rare for the thermometer to drop below 60 degrees. In Honolulu the mercury usually registers in the high 60s or low 70s at night and the high 70s to low 80s during the day.

The very feel of the air is peculiar to this trade wind zone. The atmosphere is warm and gentle, but with a vibrant freshness. It is noticeably different from that of the cool, moist air of coastal California or the brisk damp air of Puget Sound.

The visitor can more fully appreciate these distinctive characteristics of the island trade winds by leaving Honolulu, which is shielded by the coastal range, and crossing over to the north side where the lush tropical vegetation crowds down the gentle slopes to the edge of the shore. Along this northeastern coast, the wind blows onshore almost without interruption from March till mid October. Even during the period from November to February, these trade winds blow more often than not.

On the windward coast of Oahu, as on the similar coasts of Molokai, Maui and the other rugged islands in the trade wind realm, the landward moving air, thrusting up the mountainsides, generates huge layers of billowing clouds. First enveloping the peaks, they then move downward toward the coastal plains. The lands under the clouded areas are drenched, as part of the water that was accumulated over the ocean is again released as rain.

Arid sections, almost tiny Saharas, lie along the leeward side of mountainous trade wind islands. For a change from the ample rains and the lush tropical growth along the east coast of Oahu, north of Kailua, one can drive thirty miles across the island to the coast, northwest of Ewa. There cacti grow along the upper edges of the beach and are scattered along the otherwise barren slopes on the sides of the steep arroyos. Or one can bask in the bright desert sun on the Kona coast of western Hawaii. These climatic contrasts are common in regions of the trade winds, where deserts and humid portions are practically side by side, with only a mountain between.

The wind currents and the location of mountains are very important factors here. You will note on the map that the massive North Pacific Drift, a warm sea current, flows eastward, north of the Islands. This is the stream that divides when it reaches the American coast at about Puget Sound and the small branch turns northward along the Canadian and south Alaskan coasts, while the larger body turns south and in the area about opposite lower California, circles back and flows westward toward Japan. On the way it passes south of the Hawaiian Islands.

Oahu is the third largest island in size, but it houses about half the total population of the state. Honolulu, the capital, is on the lee side of this island shielded by the Koolau Range from the moist trade winds and enjoys much dry, sunny weather. There is about 25 inches of rain annually, with half of it falling from December through March, which doesn't qualify it as a rainy season. There will be about 220 days a year when there won't be even 0.01 inch of rain. Honolulu averages over 3000 hours of sunshine a year, which is about 65% of the maximum possible. The name Oahu means "gathering place" and a look at Waikiki Beach in top season might suggest that the world had taken the term literally.

Honolulu is the gateway to the islands, both from the sea and air, and many visitors never get off the island of Oahu or beyond the environs of the capital. It is understandable as the thermometer seldom strays much from the 70° low and 80° high annual average.

Between mid-December and mid-January, there can be one or several periods of unpleasant steamy, rainy days coupled with southerly breezes called Kona winds. But for about 330 days a year the dry trade winds come in from the north. There is little or no fog and thunderstorms are infrequent—perhaps five or six a year. As we have noted, the breezes are constant and rarely exceed 15 mph. With a relative humidity average of 58% and an 83° temperature, the summer city weather can be hot and uncomfortable, but it isn't many minutes drive to a higher spot for a refreshing change and even less for a dip in the sea.

You never have to travel far on the islands to find a different climate. In the case of Oahu, the prevailing trade winds come in from the northeast heavily laden with sea moisture. The high Koolau Range, which forms such a striking background for Honolulu, wrings out practically all of that moisture dropping much of it on the northeast coastal plains. Even more is precipitated out in the mountains and high country. This explains why Honolulu enjoys almost continuous sunny dry weather.

You will note from the map that while the annual rainfall averages only 16 inches in the southwestern part of the island, there is an area in the high country that gets up to 200 inches a year. The whole northeastern side of Oahu is lush green, and although only a small percentage of the island land is flat, there are many valleys and relatively smaller spaces where fine crops of sugar and pineapples flourish.

The temptation to remain in the Honolulu environs for the entire visit will be great. It has much to offer, golf courses, tennis, surfing, water skiing and skin diving as well as fine restaurants and lively night life. In spite of it all, more visitors each year are discovering that a skirmish through a few of the other islands can be rewarding.

The biggest of them all—the island of Hawaii—is 93 miles long and 78 wide, which adds up to almost as much acreage as all the other islands combined. It houses, however, only about 10 percent (70,000) of the state's total population. It is also the farthest south. South Point at the bottom tip (18°56′N), replaced Key West as the southernmost spot in the United States when Hawaii became the fiftieth state. The 13,796 foot Mauna Kea is the highest point in all the islands, and the island of Hawaii is the only place in the whole country that can boast of active volcanoes. Hilo is the capital of the island and also the orchid capital of the United States; it produces over 22,000 varieties. Unlike most of the other islands, Hawaii is in the path of prevailing winds that are from the southwest.

When the warm North Equatorial Current passes to the south of the islands on its way to Japan, the moist sea breezes blow over it and later release it as rain on the land, causing Hawaii to be known as the "Green Island." The map shows that except for the northwest corner which lies along a protected cove and gets an average annual total of 16 inches precipitation. The remainder of this island receives quantities up to 300 inches at some of the higher, cloudy spots. There can be constant misty weather; Hilo averages only 78 days a year when there isn't at least 0.01 inch of rainfall. Pepeekeo, less than ten miles north of Hilo, averages 128 inches (over 10 feet) of rain a year. The 84 percent relative humidity average is higher than the islands generally. There is no cold weather except at the mountain top, which may provide winter skiing. The thermometer has an annual spread of from 66° to 80°, and 90° will be recorded less than one day in two years. Hilo averages only 1670 hours of sunshine per year which is only 39% of maximum possible.

The island of Hawaii is the only American home of active volcanoes. Many flock in to see the molten lava streams and the even more spectacular fire displays, which visitors watch from the crater rim through long chilly nights.

HONOLULU, HAWAII — Elevation 12 Feet — Table 105

Month	Temperatures Average Max.	Average Min.	Extreme Max.	Extreme Min.	T H I*	Wind Chill Factor*	Precip. Total	Precip. Snow	Not even 0.01" precip.	More than ½" of snow	Clear	Cloudy	Thunderstorms	Fog	90° or higher	32° or lower	% of possible sunshine	Rel. Hum. A.M.	Rel. Hum. P.M.	Wind M.P.H.	Wind Direction	Storm Intensity*
Jan	79	66	83	56	73		4	0	18	0	22	9	1	0	0	0	66	76	63	10	ENE	R-2
Feb	79	66	84	57	73		3	0	16	0	18	10	1	0	0	0	65	74	59	11	ENE	R-2
March	79	66	86	58	73		3	0	18	0	22	9	1	0	0	0	71	72	59	11	ENE	R-2
April	80	68	86	60	73		1	0	18	0	20	10	0	0	0	0	71	71	60	12	ENE	R-2
May	82	70	90	65	75		1	0	20	0	21	10	0	0	0	0	72	68	59	12	ENE	R-1
June	84	72	90	65	76		T	0	19	0	24	6	0	0	0	0	74	65	52	13	ENE	R-1
July	85	73	90	69	76		T	0	18	0	27	4	0	0	1	0	76	66	52	13	ENE	R-1
Aug	85	74	91	67	77		1	0	18	0	25	6	0	0	6	0	78	69	57	13	ENE	R-1
Sept	85	73	93	68	77		1	0	18	0	26	4	0	0	8	0	76	66	52	12	ENE	R-1
Oct	84	72	92	65	76		2	0	18	0	23	8	1	0	3	0	68	68	54	11	ENE	R-2
Nov	82	70	89	61	76		2	0	17	0	21	9	1	0	0	0	61	75	63	11	ENE	R-2
Dec	79	68	85	61	74		3	0	17	0	21	10	1	0	0	0	60	77	64	11	ENE	R-2
Year	82	70	93	56			22	0	217	0	101	264	6	0	17	0	70	71	58	12	ENE	

Notes:
T Indicates "trace"
* For full explanation of (T-H-I) "Temperature Humidity Index;" "Wind Chill Factor" and "Storm Intensity," see beginning of Chapter 2.

Volcanoes have generally not been too dangerous on Hawaii, as the very slow flowing lava can be easily avoided. The acres of solidified lava beds and the glittering, jet black sand beaches such as Kalapana, however, confirm much past activity and, in the case of some particularly violent eruptions, small villages in the path of massive lava streams have been completely buried.

The prevailing northeasterly trade winds keep that side of the island moist with 64 inches annual precipitation, while the mountain-shielded drier side gets about 32 inches. But between Mt. Waialeale (5180 feet) and the equally lofty Kawaikini Peak, the average rainfall is 471 inches (39 feet) per year, the highest in the world. These torrents of water act as an immense irrigation system to the lowlands.

Lihue averages a respectable 2410 hours of sunshine a year, which is 54% of the maximum possible. There are only 161 days a year when there is not at least 0.01 inches of rainfall but this is often only as rainbow arched, misty skies. It boasts a beautiful, even climate where you will not experience one 90° day in two years.

In spite of all the slopes, many quite steep, it's an agricultural land —some rice in the swampy areas, some cattle and some pineapples, but the main crop is sugar. The visitor will also note the great variety of tropical fruits, papayas, mangos, guavas, etc., but everywhere flowers.

Maui, the second largest island is fast becoming one of the most popular with tourists. On a clear day it is possible to look 100 miles out to sea, or down 2,800 feet to the floor of the crater. Molokai, Lanai, Niihau and Kahoolawe are the other four, of the eight major islands. They are subject to much the same weather conditions and for the same reasons as those detailed above.

Hawaii is generally considered one of the most healthful areas. There need be no concern about eating salads, raw fruits or vegetables. Drinking water is excellent—and so is the rum!

We know of no other part of the United States (or elsewhere) that has more to offer weatherwise. Information can be obtained from the Honolulu Board of Realtors, International Savings and Loan Building, Honolulu, Hawaii, 96815. Also from either of the two offices of the Hawaii Visitors Bureau at 2270 Kalakaua Avenue, Honolulu, Hawaii, 96815 or 609 Fifth Avenue, New York, New York, 10017.

5

Retirement Spots

See Chart No. 38

More and more people now look forward to retirement in pleasant anticipation rather than with resignation and perhaps a tinge of apprehension. In most cases, age is no longer a barrier to travel and other enjoyable activities. It is, however, advisable to start thinking well in advance about the things you would like to do and where you will find the most agreeable living conditions.

Unfortunately, many people equate a pleasant vacation experience with the ideal permanent abode. A few weeks of wall-to-wall sunshine in the Arizona desert country could convince a Puget Sound "webfoot" that there indeed is heaven, but three hundred and sixty-five somewhat desiccating days could also change his mind. So, too, a New Englander shivering in the wintery blasts might hark back nostalgically to that balmy December in southern Florida. To many fugitives from the northern climes, Florida is indeed a perfect year around home. Others find the summer heat and humidity oppressive, the relative uniformity rather monotonous, and long for the distinct change of seasons. If one swallow doesn't make a summer, neither does one pleasant vacation make a retirement utopia. For retirement, we must be concerned with the year-round weather conditions.

Retirement almost always involves some radical changes in way of living but also often permits considerable freedom of choice in location. All too few prepare for retirement, which often results in great disappointment and sometimes real tragedy. A retiree who immediately sells his home and rushing off to that promised land buys another, may find to his great sorrow that this new location is not the perfect year around spot.

Of the many things which a prospective retiree might do well to ponder, there are two which seem particularly important:

1. *Do as much research as practical* over a period of time. Chambers of Commerce, state and municipal bureaus, guides, magazines and often newspaper items can all be most useful. Several of our friends, contemplating retirement, subscribed to newspapers in the areas which seemed most interesting. This not only gave them a good idea of the people, facilities and types of activities, but allowed them to make day-to-day comparisons between the weather there and in their home area. This we highly recommend, as the weather figures by themselves can be rather meaningless. Comparing them with those you are actually experiencing, however, should be very enlightening indeed. Most complete and detailed weather information is available from the U.S. Government Printing Office in Washington, D.C., but unfortunately, much of it is couched in meteorological terms. Hopefully this book will translate that semi-technical language into an everyday story of weather.

2. *There is no perfect spot* but by the process of elimination you will arrive at the place which most nearly coincides with your tastes. Visit this area of interest—several times if possible, and in different seasons. Even a short stay will be most useful for gathering much on-the-spot information and there is no substitute for actually seeing for yourself. But all of this will certainly not be enough to justify making an immediate final decision. By far the best procedure is to live there on a temporary basis for a year, if practical, before purchasing property.

While a general area may be very pleasant there can be less satisfactory mini or sub-climates which might dictate the selection or exclusion of a particular section. For example, in the delightful town of La Jolla in southern California, there are districts where when you cross definite lines some mornings or evenings, you can emerge from thick fog into the brightest of sunshine. Even in Buffalo, New York, the winter streets in the south end can be quite bare while cars are stalled in twenty inches of snow at the north end of town. You may not end up in either of these spots, but similar phenomena prevail in many places, and you seldom learn of them from the literature or casual conversation. The most reliable method is always on-the-spot investigation.

One important thing to remember is that there are at least as many

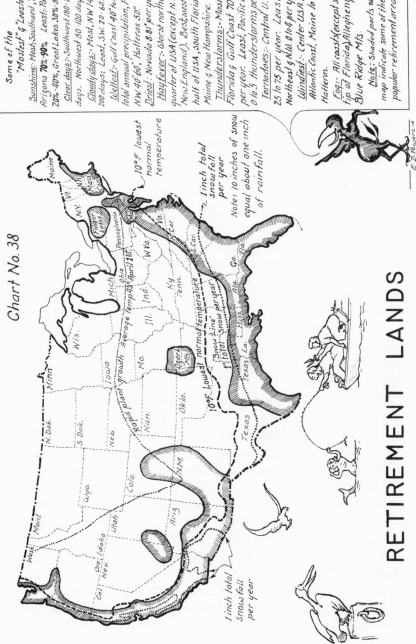

Chart No. 38

RETIREMENT LANDS

tastes regarding weather as in the selection of neckties. For example, as a relatively new resident in South Africa, I was rather disappointed to wake up and find an overcast morning, after eight months of glorious sunshine and cloudless skies. The reaction of an English friend, however, came as a bit of a surprise: "Lovely gray day, isn't it?"

Years ago, most retired persons continued to live in their old homes or at least in the same general area. While this is still true to a major extent, increasing numbers are now migrating to spots that seem to offer a more pleasant and comfortable year–round way of life. For those blessed with good health this can mean greater opportunity for outdoor activities. Many seek gentler climates which they hope may bring relief from physical ailments.

We strongly urge, however, that you obtain medical advice before moving to a markedly different type of climate. This applies to those who think they're robust as well as those who know they're not. Your family doctor or specialist most probably won't be familiar with the exact weather conditions in the place you are considering, but if you have done your home work, you can describe how conditions compare with those of the area in which you are currently living. This can form quite a sound basis for his evaluation and recommendations.

Doctors generally agree that a reasonably uniform, mild year–round climate is most suitable for people over sixty years of age. This doesn't mean that you can't live to be ninety-five in the stormy Great Lakes region, the wet Puget Sound country, or the Kansas flats. They just think the odds may be better.

Many doctors suggest the south just as a general precaution to avoid the possible strains imposed by severe or stormy weather. There seems to be no absolute evidence that there is any connection between heart attacks and the violent northern weather, but statistics suggest that people suffering heart trouble or arteriosclerosis appear to do better in the milder climes.

More and more attention is being devoted to "climatotherapy." This is the treatment of disease by subjecting patients to an appropriate climate. Physicians generally agree that climate, while important to health particularly for the elderly, is only one factor in promoting overall well-being.

Whether you stay or move after retirement can be influenced by many considerations. The following check list of eight items will be of interest to most retirees.

1. Which climate seems best for your particular condition? This is especially important for those with physical ailments. Check the available medical facilities.

2. It's becoming increasingly important for elderly people to avoid a contaminated atmosphere. The culprit may be smoke, smog, dust, industrial pollution or excessive auto fumes. We suggest that you read the section on "Smog and Air Pollution" in chapter 1.

There are a million allergies—some of which can be alleviated by omitting eggs from the diet or staying fifty feet away from dogs, but the one we are concerned with here is pollen. The severity varies greatly in different areas—some of which are quite close together. One thing that may come as a surprise to many is that hay fever season is not the same in all parts of the country. This subject is covered in some detail under "Hay Fever" in chapter 1. Medical men, hospitals or newspapers in the area can usually also be of assistance in this matter.

3. Which area or community will permit the greatest opportunities for physical recreation, preferably out-of-doors? What are the geographic features—mountains, highlands, lakes, sea coast, etc.—which have the most appeal to you?

4. Many find that being close to a college or university almost guarantees a wide scope of cultural and entertainment activities, and general mental stimulation. Other factors are resident ball clubs and performing arts.

5. Observe the type of people in the area, the degree of affluence. Are there other retirees; what are the opportunities for social participation?

6. Cost of living, foods, types and cost of houses, apartments, or other accommodation.

7. Are part-time jobs available, opportunities for a small business, or capitalizing on a hobby?

8. Is the location too isolated or is it reasonably close to a larger city for an occasional visit or medical and dental attention? For some, it's important to be able to get back home once or several times a year. Even when they don't often take advantage of this convenience, the fact that they can do so removes much of the sense of distance and isolation.

Several of the above items and no doubt many additional ones will be of considerable interest to a certain number of people, but only two —"climate for the retiree" and "air pollution"—will be of great concern to all, and it is to these very important subjects that we will devote the rest of this chapter.

249

What Are the Choices?

By far, the greatest number of persons who move after retirement choose a spot in the relatively snow-free areas. This includes the California coast, Florida, the whole Gulf crescent, sun drenched southern Arizona and New Mexico, the eastern part of Georgia and South Carolina. To the surprise of many, the Puget Sound and the coastal Pacific Northwest also qualify in this category. A number of major retirement areas are shown on chart no. 39. In this vast area comprising portions of a dozen states and over 2500 miles of sea coast, there is one great common denominator—a soft mild climate and a minimum of cold stormy weather and snowfall.

As might be expected, there are also great differences, particularly in the amount and periods of greatest precipitation. In some sections summer is the wet season; in others it is winter. Florida and the Gulf get 70 to 90 thunderstorms per year, while the California coast hardly sees one. Neither is the West Coast much threatened by hurricanes, but it does have frequent heavy fogs which when coupled with pollution produce poisonous smogs.

While New Orleans has a rather high degree of air pollution, aside from one spot in the Florida panhandle and a few areas along the Texas coast, the Gulf area is rated fair to quite good from the viewpoint of pollen pollution.

Some of the other less publicized retirement centers are equally attractive to many.

The Ozark hills region—which includes northwestern Arkansas, southwestern Missouri, and a bit of the northeastern part of Oklahoma—has long been the happy playground of vacationists, and is fast becoming equally popular with retirees. A pleasant and inviting island in a vast sea of flat lands, it has a continental climate characterized by a distinct change of seasons. Winters are brisk but not too cold, and summers are warm to comfortably hot. There is a moderate 45 inches of precipitation which is fairly uniform throughout the year and falls mostly as rain. As an example, Little Rock averages only 5 inches of snow a year, which may cover the ground a total of 15 or 20 days. There is very little freezing weather. It enjoys a high 62 percent of maximum possible sunshine. There may be fog about 15 days a year, and you may also expect 60 thunderstorms annually. The 57 percent average humidity is remarkably uniform throughout the year. The Ozarks, Ouachita

250

and Boston highlands area is made up of rolling country, hills and mountains with deep valleys. There are many small communities nestled down in those valleys that enjoy remarkably mild winters, being protected from severe blasts by the surrounding mountain ranges.

The "research triangle" of North Carolina attracts many from the arts, particularly writers. But the newcomers are also retired doctors, professors, and other former professional people. This is a community of culture, leisurely living and a soft mild climate. The sometimes high mid-summer humidity can be a bit oppressive, but residents point out that they are only a couple of hours from either the cool mountains or sea breezes. Of the three communities cornering the triangle, Durham (where the outstanding Duke Medical Center is located) is perhaps the least attractive. Raleigh, the state capital, is a most pleasant southern city, and Chapel Hill a delight. The year–round climate in this attractive rolling Piedmont country is very agreeable. For instance, even though Raleigh is 430 feet above sea level, records show only about one snowfall a season. It has a high and low temperature range of 89° to 32°, with a few days up to perhaps 104° and some winter nights down to 7°, but quite uniform on a day to day basis. The total 45 inches of precipitation is also quite evenly distributed but a little heavier during summer. There will be 45 to 50 short thunderstorms from May through September. The humidity which is remarkably even, averages a good 53 percent. If you want a quiet, comfortable way of life, in pleasant cultural surroundings, and at quite reasonable costs, this generally unpublicized part of the country may be of interest.

The entire state of North Carolina has much to offer. Just to the south is the famous golfing country of Southern Pines and Pinehurst, one of the most pleasant retirement areas of this country. Living costs can be higher than many parts of the state but there is also the usual range available. The weather is much the same as the Raleigh area, though a bit milder.

Charlotte, only 100 miles to the west but at 725' above sea level, almost duplicates Raleigh's weather except that the thermometer may drop to −3° a few nights in January, and there will be a little more sunshine and a bit less fog. The added elevation also means generally clearer air.

Both the Great Smoky and Blue Ridge mountain country to the west and the low coastal plains also satisfy many retirees. Asheville is reasonably typical of the high lands, enjoying much milder winters than might be expected. Although the thermometer can drop to 9° on the rare occasion, the average range is low of 29° in January to 85° in August. The clear dry mountain air, together with almost 60% of the maximum possible sunshine, is most agreeable. There are about 45 days of fog and 50 short thunderstorms a year. Total precipitation is 37 inches, quite uniformly distributed throughout the year, of which the 12 inches of snowfall is of the dry, light type.

Wilmington, in the southeastern part of the state, is reasonably typical of the low coastal flats and the ocean front. This waterland country is slowly beginning to rival the Chesapeake Bay region in popularity and offers much the same wide range of marine and beach activities. It also, of course, enjoys a considerable weather advantage. The temperature averages from 74° down to 53° with only the occasional plus 100° day and 10° night. Fog incidence is a low perhaps 18 days per year but there are about 45 thunderstorms, mostly in the summer. Precipitation totals 50 inches, being heaviest from July through September, and includes only 2 inches total of snow. Humidity is quite uniform throughout the year, being a good annual average of 57 percent. This eastern point of North Carolina may be a bit more in the path of hurricanes than most of the coast, but the incidence of these storms is fortunately few and far between.

Various sections along the whole Atlantic seaboard and a hundred miles inland are also being more favored by retirees. The southern half of the Maine coast, the Cape Cod peninsula, the New York Finger Lake district, and similar sections of New Jersey and Pennsylvania all have their champions. In most of these there are hot, sunny and generally pleasant summers with some intervals of high humidity. The northeastern sector enjoys the most glorious autumn in the world, but suffers with damp, blustery winters, and inland and higher portions get heavy snowfalls. Along the coast, rain often predominates. There may also be delightful short spring or just as often, none at all.

Many choose the Chesapeake Bay shores as being an in-between climate. While reasonably free from extremes, there is considerable damp weather, some not too unpleasant but often bone chilling. To compensate, the whole region is a marine playground, most often enjoying bright pleasant summer and fall seasons.

Living costs can be very high, as the Eastern Shore has long been a prime choice of the wealthy. All through the Del Mar peninsula, however, there are many pleasant low-cost communities both inland and near the water. Baltimore, at the head of the Bay, is reasonably representative of this section. While it may have the occasional below zero night, winters are short and (although wet), not severe. The 43 inches of total precipitation is distributed almost equally throughout the year, with 3 or 4 inches each month. The percentage of maximum possible sunshine is 57% and the 32 days of fog per year can be expected near the water. The average and quite uniform relative humidity is a healthy 55% For those who enjoy fishing, crabbing, sailing or other water activities, this might well be the ideal spot.

Much of Virginia, such as the Albemarle country around Charlottesville, offers a pleasant way of life with a moderate climate. Although this is north of what is loosely called the "snow line," there is not much severe weather. Located in the Blue Ridge foothills, it is generally a crisp invigorating winter place. Probably what would disturb most, are the few weeks of hot summer days. This is historic country and the University of Virginia is a center for lectures, music and cultural pursuits, as well as, for sports and other outdoor activities.

All the way down to the North Carolina border is wonderful retirement country. Even such places as Wytheville, in the southwest corner and perched up at an elevation of 2300 ft. above sea level, enjoy a pleasant mild winter with temperatures seldom dropping below 25°. The uniform precipitation adds up to a low 28 inches a year. For a most agreeable leisurely way of life, with distinct seasonal changes and almost no severe weather, this whole stretch through the center of Virginia and the Carolinas is proving to be the answer for more retirees every year.

The two strongest weather magnets—for both vacationers and retirees—have long been Florida and California.

Florida

We think of Florida as being a Southern state. However, if you divide the state with a line drawn from Jacksonville through Gainsville to perhaps Cedar Key, you will find two quite different types of country —both from the viewpoint of climate and atmosphere. Above that line, including the whole panhandle, the country fits what we think of as the Deep South, with a Gulf Coast climate. This upper area will be included with that long coastal sweep, later in this chapter. South of that line is better characterized as Northern oriented, with a mild to sub-tropical climate.

A wonderful climate is Florida's greatest plus factor. The average mid-winter temperature in the northern portion of this area is about 55°, and 68° in the southern section. Conversely the summer temperatures are generally no higher than those in New York City or Chicago. While it can't boast Arizona's guaranteed sunshine, the sun shines 5 or 6 hours a day in winter, while most of the North averages less than 3 hours. Nor is it much troubled with fog, smog or air pollution. In recent years, however, excessive automobile exhaust fumes have become a serious consideration, and it's well to avoid heavily trafficked highways and areas of high auto density.

Florida does have its climatic excitement—sometimes with a touch of violence. It averages over 50" of rainfall a year, with half of that falling from June to October and perhaps 15" from September to November. Winters are relatively dry with only about 6" during the whole season.

This state also experiences more thunderstorms than any other part of the country. The Tampa section is the center of greatest activity and has about 90 per year, while the southern tip and northern panhandle get about 70 to 80. More serious are the infrequent hurricanes. These were quite frequent during the 1940s but only two marginal hurricanes passed over the state from 1951 through 1959. Although the physical damage can be severe, the excellent warning system eliminates much of the personal hazard. All in all Florida is a fine place to live, and living costs can be very high or quite modest as you choose.

Key West is the southernmost spot in the forty-eight states. If it were moved across the same latitude to Africa, you would have to travel some 400 miles north to get to Cairo, Egypt. Key West is about 500 miles south of San Diego, which is almost at the Mexican border.

Key West does indeed have a beautiful climate—never a hint of frost and a temperature range of only 25° throughout the year. With an August average temperature of 88° high and a January low of 65°, and the usual sea breezes—living can be very pleasant. The total annual precipitation is a low 39 inches (no snow, of course) and falls mostly as showers, sometimes in the company of the 60 short thunderstorms which can be expected each year. The sun shines 68% of the maximum possible during daylight hours. If you can afford its high cost of living there are few better spots to fish, swim and loaf in the delightful informal atmosphere among the many writers, artists, retired naval and military personnel and professional people.

The 100-mile chain of islands, joined to the Florida mainland by an excellent auto causeway, enjoys much the same conditions. This is true of climate, cost of living, availability of living space, and outdoor activities.

Marathon, in the center, is an active outdoor community with plenty of fishing, sailing, and boating. At the upper end are Key Largo, Tavernier and Islamorada. Finding suitable accommodations may be difficult, since both rentals and places for sale are scarce. We recommend that you do your looking in other than the high winter season and the July, August school vacation time when the area is crowded.

Miami and environs. Traveling northward from the Keys, your first stop might be Homestead—a flattish, inland very productive agricultural country where many retirees have begun to settle. The attractions are quiet, low costs, and plenty of sunshine.

The Miami area is certainly the center of tourism—winter and summer. Because of high costs, noise and hectic activity, few retired people remain in Miami Beach proper unless they are associated in some way with the tourist business. The beach strip does enjoy, to a greater extent, the pleasant sea breezes.

Miami Shores, only a few miles north of downtown Miami, is a pleasant community of quiet, tree shaded streets and neat single-family homes. The prevailing easterly sea breezes, which help greatly to relieve the mid-summer high humidity, do not penetrate to all of these inland spots. For example, while Miami Beach averages less than 18 summer days a year over 90°, the city of Miami proper, which is on the mainland, will have over 65 such days. On the other hand, this whole east coast strip of Florida experiences about the same number of summer days over 90° as does New York City—usually 15 to 20.

A short distance southwest of Miami is the charming city of Coral Gables. A center of culture, it is a more expensive section, with most homes in nicely landscaped settings. Strict zoning, with practically no tourist facilities, has maintained the refined and well-planned appearance of homes and streets. A person can live a quiet, comfortable life in this center of cultural activities and still be close to the recreational entertainment of Miami proper and the Beach. The average mid-winter temperature is about 68°, some 10° higher than the Los Angeles area. Summers are comfortable with a mean of 80°. Like most of the general surroundings, Coral Gables enjoys plenty of sunshine and the annual rainfall is 55 inches, mostly in short summer showers. It is relatively free of excessive auto exhaust fumes.

About 15 miles north of Miami is the more resort type Hollywood. The coast is lined with hotels, motels, restaurants, and all manner of tourist facility, but the inland has been taken over by a mixture of retirees, active residents, and more tourists.

Further up the coast Fort Lauderdale, with over 100 miles of canals, rivers and an immense marina, is a delightful center of boating, fishing and water life.

It is an interesting phenomenon of most any high grade spot in Florida that retirees, sometimes after much searching, can find anything from the very tops to most modest accommodations. It is easy to understand why this beautiful, water garden city is so popular. The climate is similar to that of Miami—never too cold and the usual sea breezes relieve the hot spells.

Between Miami and the socially-oriented Palm Beach are Boca Raton, the smaller, less glittering Pompano, Delray, Boynton Beaches, and other still smaller settlements. There is ample space here for retirees. The same is true all the way up the coast, past St. Augustine, the oldest city in the country, to busy Jacksonville. It's a bit cooler here, but its average winter low is only 45°. Although there is seldom more than a trace of snow, this portion of Florida is really more of a summer than winter resort. There will be the odd below-freezing nights about 5 months of the year and the 35 fogs per year occur mostly in winter. The 52 inches of rain is spread throughout the year but is heaviest during summer. It is cooler here and a favorite spot of many retirees.

About halfway up this stretch of eastern Florida coast is the famous beach at Daytona. A pleasant average high winter temperature of 75° with a low of 50° and 63 percent of the maximum possible sunshine hours, make this a prime winter resort. It is equally popular in summer, with a quite uniform relative humidity of 55 percent throughout the year. There are frequent but short afternoon showers and usually a sea breeze. As is true in much of the south Atlantic and Gulf coastal regions, many of the summer visitors are from inland parts of the southern and border states. Living styles range from ocean-front estates to reasonably modest homes and, of course, the full complement of luxury hotels to standard motels.

The Florida West Coast, bordering the Gulf, enjoys a more relaxed and quieter way of life and perhaps attracts more older people both as visitors and residents. Here the sun shines a higher percentage of the time than on the more publicized "Gold Coast" on the Atlantic side.

St. Petersburg and the surrounding districts form the largest retirement center in Florida. Many restaurants, entertainment, recreational and other activities, such as the several hundred shuffleboard courts, are geared to the aged and in many cases, the feeble and invalids. This is evident from the well-occupied famous green benches to the gentle ramps in place of steps. It would be difficult to find a more all around satisfactory spot. The climate is just about ideal. The sun shines about six hours a day in winter, and the *St. Petersburg Independent* gives away free copies of that newspaper on the 5 or 6 days each year that the sun fails to shine by press time! About 80 percent of the 50 inches of rainfall occurs in other than the winter months during which season the mean high temperature is about 70° and the low 50°. The summer thermometer ranges between 70° and 90°. Many doctors recommend this as one of the healthiest places in this country for older persons.

Clearwater, just up the Pinellas peninsula, (and smaller Dunedin, fifty feet above the water level) is one of the most popular and healthiest spots in Florida. Living costs may be somewhat higher than in the general area. North along the coast past Tarpon Springs (originally settled by Greek sponge fishermen) and up to Yankeetown it is sparsely settled, but there are many fishing settlements and camps. From Cedar Key, in Suwannee River country, north including the entire panhandle region, the atmosphere is that of the Deep South and the climate typically that of the Gulf Coast. That area will be included later in this chapter.

The area from St. Petersburg south to Everglades Park has probably developed more in the last 15 years than in the previous century. There are many delightful communities large and small, beautifully landscaped with lush sub-tropical vegetation. The Spanish-style architecture adds to the soft charm, while zoning and other regulations serve to maintain the atmosphere of pleasant, comfortable living.

Sarasota is the largest of this chain. Venice, Fort Meyers, Naples and a number of others are all charming, well regulated communities. Most of the choice waterfront properties indicate an upper-income bracket population. You will find little of the glitter and bustle so evident along the Atlantic side, but have no fear of stagnating. There is plenty of activity—cultural, outdoor and social—it is simply keyed to a lower tempo. The new Alligator Alley Road will make an east coast visit a short comfortable jaunt. You should know that it can be unpleasant in spots when the very infrequent "red tide" occurs. This is produced by millions of minute red marine organisms that suddenly appear, discoloring the water and causing the death of many fish. If you are familiar with the fragrance of a week old fish in hot weather, multiply that a lot of times. Luckily it doesn't happen often.

It should be remembered that the prevailing winds are generally easterly across the land. This, in a measure, accounts for the 120 days a year of over 90° temperature, which Fort Myers usually experiences. Residents attribute the insect density, which appears at times during the year, to these same land air currents.

Central Florida—The higher inland elevation at Gainsville, along with the somewhat less than average 55 inch Florida rainfall and lower humidity, seems to favor folks with bronchial problems. The slightly lower winter temperature (maximum high average 70° and low 45°) is apparently not a serious factor. Many also find the bracing climate of the Winter Park vicinity generally beneficial, and it has become one of the most popular retirement areas in the United States. With Rollins College as a hub of cultural activities, Winter Park attracts artists, writers, professional people and those who enjoy music, the theater, and lectures. In addition, there are the usual outdoor and recreational functions.

Orlando at 100 feet altitude and Lakeland at over 200 feet, are both preferred to places on the Gulf or Ocean by some retirees who find the degree of humidity more to their liking. Although the elevations in this central section of the state are rather modest, they often catch summer breezes missed by the lower inland places.

We have covered only a few of the many desirable places in Florida but have tried to include some of each type in all sections of the state. Hopefully this will serve as a basis for you to start your own research.

We urge that you investigate, and if at all possible, rent for a while, before buying a home. There may be a weather or other condition which you will find intolerable. We've known people who were all settled in their newly purchased home before they discovered with dismay that all of those nice beaches and fishing spots were "private." The Florida News Bureau, c/o Florida Development Commission, Tallahassee, Florida, 32304 is an excellent source of information.

Gulf Coast

This immense expanse of mostly sub-tropical country is one of the three major retirement areas in the United States (the other two being South Florida and even more popular California.) By and large this has the lowest cost of living index of the three, and in some parts, is about as low as you will find in the entire country. For those who can tolerate the sometimes-extreme humidity, this can be a retirement zone hard to beat.

Winters are mild but not hot. As indicated in the tables and map, snow is rarely seen and then only for a few hours. January average mean temperature is 52° but the many bright, sunny days make it seem warmer. The whole center segment of the Gulf Coast, from eastern Louisiana to the west end of the Florida panhandle, gets an average rainfall of 60 inches a year. The only other large areas that equal this heavy downpour are the Pacific northwest and most of Florida. The rain pattern decreases progressively as you go westward from Louisiana until you reach Brownsville, Texas, which gets 27 inches per year. The summer average mean temperature of only 84 degrees can be quite humid. The thermometer seldom drops below freezing in most of this area, but the 90° days can begin about the middle of June. New Orleans and Mobile average 70 of these high temperature days per year. Brownsville gets the high 115 of such days but Pensacola only 39. An interesting example of the effect of off-water breezes is Galveston which averages only 35 days over 90°, while Houston, hardly 50 miles inland suffers with 90 such days. Because the off-Gulf breezes are such an important comfort factor, you will note that some spots close to the water may be much more agreeable during this period than others back inland. Pensacola or Corpus Christi can be more comfortable than Mobile or New Orleans if there is an off-water breeze on a day of very high humidity.

There is, of course, always the possibility of hurricanes, but the overall records show that there appears to have been a shift in direction over the years. From 1900 to 1938 a greater number hit the Gulf Coast, while from 1938 to 1964 more have turned towards the East and swept up along the Atlantic. But these storms are most unpredictable and some observers believe that the pattern is again moving toward the west. Hurricane Camille, which centered on the Gulfport–Pass Christian sector, caused severe damage in a wide sweep and was the major natural catastrophe in the year 1969. Fortunately, hurricanes rarely hit land and the excellent government warning system eliminates most personal injury and greatly reduces property damage.

For those who enjoy fishing, boating, swimming and other water activities, the Gulf coast is ideal. There are also excellent facilities for golf and most any other outdoor action at which a retiree may wish to flex his muscles. And for lovers of seafood, this is a gourmet's paradise. There is also much of cultural and historic interest in this scenic land.

There are a few drawbacks which should be brought to the attention of prospective residents. Annoying insects may be over plentiful at times, but selecting a place favored by off-water breezes can sometimes be the answer. The insect problem may be found in most any part of the country, but one which is peculiar to this general region is the strong, sickly sweet smell of the paper pulp mill.

You may be subjected to this annoyance only when the wind is from a certain direction so it is well to inquire if there are any such mills in the neighborhood.

While it is generally true of this whole part of the country, Pensacola in particular has a plentiful supply of crystal clear, pure underground water. That, together with a magnificent climate and living conditions, has attracted considerable industry.

The Georgia shores are washed by the mighty Atlantic but are only a few miles distant from the warm soothing waters of the Gulf Stream. This state could properly have been included with North and South Carolina but is perhaps a bit more akin to its Gulf coast neighbors. Georgia's mild climate makes it winter resort country.

Savannah's semi-tropical summer (average high 90° and low 70°) is made more pleasant by the five or six mph off-sea breezes. Winters are mild but there can be some damp, raw days and fog perhaps twenty times a year. The annual 46 inches of rainfall may seem like more, as most comes in heavy showers during June, July, August and September and there is a trace of snow during the four cold months.

Perhaps more a vacation than retirement area is the famous Sea Island some 100 miles to the south. Although it has long been a playground of the wealthy, many retired folks of modest means have settled in this vicinity.

Atlanta, having an altitude of 975 feet and being less damp, gets an inch or two of snow a year. The winter air is crisp and invigorating; the thermometer will drop to 36° and occasionally 8 or 10°. Being in the lee of the protecting Blue Ridge Mountains provides shelter from most stormy blasts. The hottest mid-summer monthly average is 89° and the July average minimum is 70°. The relative humidity averages 55 percent year round. Atlanta is the political, business, cultural and educational center of the state. There are numerous sporting facilities, both active and spectator. Many have taken advantage of the new, quite reasonably priced apartments.

Columbus and Albany in the lower, flatter country, further south enjoy a little lower cost of living than the capital. Other favored city areas are Macon, Augusta, La Range (a delightful place), Rome, Marietta and Brunswick. This whole area escapes any really severe weather and still enjoys the variety of a distinct change of season.

There is a string of new and very pleasant communities along the Florida panhandle which are attracting a great number of people. In appearance, temperment and climate, the Panhandle country much more nearly resembles the Gulf Coast than it does the southern part of Florida. From Cedar Point (a quiet little fishing spot) clear around to Panama City, there are only a few small towns but a great number of fishing settlements and camps. Most of this water edge land is swampy. There is not much here that would appeal to most retirees as a permanent home but it's a wonderful fishing and vacationland for those living within visiting distance.

The charming capital city of Tallahassee has all the attractions of the deep south. It's also close enough to the coast to take advantage of most water activities. (Write to the Florida News Bureau c/o the Florida State Development Commission, Tallahassee, Florida, 32304.)

Pensacola, Panama City, Fort Walton Beach, and Destin are all wonderful places for people who love sunshine, water activities, and everything else we said about southern Florida. The winters here are cooler and the thermometer can drop on occasion to 30° or lower. You must also expect an occasional spell of disagreeable weather. The average January temperature is about 55° as against the 66° Miami area. Summers, however, are a little more pleasant. There is one 25 mile stretch of pure white sand beach as fine as you can find anywhere in the world.

Alabama, Mississippi, and Louisiana are known as the land of azaleas and camellias. No less satisfying are the crabs, shrimp, red snapper, pompano, etc. Unlike southern Florida where a great number of the beaches are "private," most of this entire coast is free and open.

261

Mobile has all the friendliness, courtesy and pleasant atmosphere of a small town but the facilities and services of a metropolitan center. Situated back from the Gulf and to a degree blocked from the breezes, the city proper suffers from extremely high summer humidity. While few retirees might choose to live in the immediate city, there are, however, many pleasant suburban communities. The spring is gorgeous, the winter generally pleasant, as too is most of the fall season. Point Clear, with the Grand Hotel at the tip, is rather expensive but Fairhope, perched atop the 60 foot bluff, also just across the bay from Mobile and noted as a health center, is one of the most popular spots for retirees of more modest means. So too are Daphne and Montrose just to the south, but the short coastline of Alabama, as well as considerable distance inland, is becoming a most favored area for northern retirees.

The equivalent Mississippi country is quite similar. The unfortunate 1969 hurricane disaster which zeroed in on the Gulfport–Pass Christian stretch wiped out many small retirement homes. Fortunately, hurricanes occur very infrequently.

Biloxi is a bit on the honky tonk style but there are some very attractive spots in the immediate vicinity. Across the bay, Ocean Springs with its mineral springs is a popular health center.

Another very celebrated health area, just across the Louisiana border, is known as the Ozone Belt. This includes the pine-forested lands across Lake Pontchartrain from New Orleans. Here the soft climate, fragrant pine-scented air and the pure artesian wells of crystal clear water are said to have therapeutic properties which attract convalescents, invalids and many very robust people. Webster defined "climatotherapy" as the treatment of diseases by subjecting patients to appropriate climate. Some medical men suggest this place for those suffering from respiratory problems, heart disease and other ailments. The results have satisfied many. It's a pleasant, quiet retirement atmosphere but only an hour from New Orleans.

Tiny Abita Springs, Sidell, Hammond and other parts of Tammany Parish and adjacent areas are all attractive for retirement. Going west to the Acadian Evangeline country of Lafayette, New Iberia and St. Martinville, you pass through the active oil town of Morgan City. This quiet section of lower Louisiana will no doubt become a popular retirement community in the near future. It has a charming Cajun atmosphere, augmented by savory dishes peculiar to the region. Here you would certainly want to rent before you buy, as some will love everything about this back water country—others wouldn't care for it at all.

There are a number of quiet, different but equally attractive communities in the Texas sector of this long sweep of coastal territory. Austin, the capital, while overshadowed by the giants Houston and Dallas, offers just about everything a retiree or anyone else could wish for. Being a bit inland, and at an elevation of 615 feet above sea level, the air is drier than at the coast, with a 35-inch annual rainfall, and an average relative humidity of 75 percent. The summer thermometer will range from about 70° to 85° and the lowest monthly winter is about 40°. Winters are not severe, with temperatures seldom as low as freezing. The sun shines 65 percent of the maximum possible hours.

Living costs in Austin are higher than most of the Gulf Coast but lower than the other large Texas cities. The quiet and climatically agreeable hill country nearby and large lakes make fine retirement spots.

San Antonio is one of the most attractive and colorful communities in the whole United States. It is one of the few places where the river, which winds through the city, has been landscaped for park and recreational use. In addition, there are over 2000 acres of park lands. Much of the city has retained a pleasant Spanish atmosphere. There are a number of small universities and junior colleges. San Antonio is fast becoming one of the most important retirement centers in the U.S. The climate is quite similar to that of Austin, with only a bit over 25 inches of rainfall and somewhat more sunshine—fifty percent of the possible in winter and 70 percent in summer.

Corpus Christi is a big, bright, sun-drenched oil town. Clean, prosperous and a bit expensive. It's a marvelous place for active fishermen, also one of the best for bird watching—particularly near Port Aransas. An agricultural and cattle country dotted with oil derricks, it is clean, prosperous, and a bit expensive.

Brownsville gets you a little way from things. Many seem satisfied with its marvelous weather, only 27 inches of rain, lots of sunshine, a pleasant yearound climate with a summer high of 94° and a low of 75°. Brownsville averages about 115 days 90° or higher a year. Practically every day during July and August reaches that figure. The humidity can be high and if not relieved by the cool Gulf breeze, summer days may be very uncomfortable. Winter temperatures range from 70° down to 50°.

California

As the third largest state, with over 800 miles of coast line, several lofty mountain ranges and one of the world's major deserts, there's plenty of space in California for a variety of climates and an equally wide range of tastes. Most retirees choose the southern half of the state and are pouring in at an alarming rate. One of the most attractive features of this area is that no matter which type climate you select, other distinctly different locales are within easy visiting distance.

You are probably familiar with many of California's climatic assets, such as the plentiful sunshine, soft sea breezes, mild winters and pleasant way of life. But it's not always such; overcast skies, fogs and storms do occur. Then, of course, there are the dense fogs along the coast which become smog when polluted with noxious contaminants. But this strip practically never experiences a thunderstorm. And so it goes, each area has its plus and minus weather factors.

One natural phenomenon that many incorrectly associate almost exclusively with California is the earthquake potential. The San Andreas fault, which is a long fracture in the earth's surface extending almost from San Francisco to San Diego, has been widely publicized even though there are several others in the state as well as many throughout the country. It has been predicted by several scientists that sometime in the indeterminate future, the lock causing the build-up of internal strains will release causing a catastrophic earthquake.

Let's start on the coast just above the Mexican border, an area which so many retirees have found to their liking. From there, north to Los Angeles is an almost continuous chain of communities, the largest by far being San Diego. The major attraction of this beautiful coastal strip is, of course, the blue Pacific and a most pleasant year-round climate with practically no extremes of temperature.

Using San Diego as an example, the thermometer will range from a summer high of about 78° down to a January low of 45°. The total annual rainfall (there is only a very infrequent trace of snow) of 11 inches falls during the winter months. The scant three days a year when the thermometer registers 90° do not qualify as heat waves. Off-sea breezes during nine months of the year are refreshing, while the prevailing southeast winds in December, January and February reach this section after passing over the warm desert country. Few places in the world will come much closer to a climatic "utopia" than southern California.

For the many who prefer smaller communitiès there is a wide choice, all enjoying much the same weather conditions. One of the most desirable is the charming little La Jolla, which is just about ideal if you can find accommodations to suit your purse. There are many other lovely little places, somewhat lower in the cost of living index, which are worth investigating. Unfortunately, in recent years, particularly when there is the infrequent northerly wind, smog envelops most of this area for short periods. There are often night time fogs which, however, usually burn away by about 10:00 A.M. It is interesting that the fog follows quite definite patterns; you may emerge from a dense fog bank into bright sunshine. Only on-the-spot observation will identify these patterns.

Residents in the San Diego environs need never want for things to see or do. There is a high degree of culture, many places to visit and a great variety of outdoor activities within easy reach. It's a short drive into Mexico, the desert country to the east, or to the mountains and several National Parks. The carpeting of spring flowers on the desert when the temperature is most agreeable, and the glorious display of color in December, just before the Del Mar and Encinitas growers have shipped out the poinsettia crop, are sights not to be missed.

An inland community less than ten miles from the coast is the delightful little settlement of Rancho Santa Fe, which over the years has remained almost entirely residential, excluding all but most essential shops. Here it is a bit warmer, but a fine year-round climate.

As you approach Los Angeles, the population density increases as do the cost of living and smog. It is unlikely that the immediate Los Angeles area will have much appeal to most retirees, but just to the coastal north are pleasant communities, large and small, not too different in climate from places further south.

Santa Barbara is a place of exceptional beauty and charm which has long been popular for year-round living. By custom, there is an off-season rate in tourist accommodations from September till June, but there is no off season for fine weather. Like along most of the coast, the rain (18 inches per year) occurs mostly during winter. The temperature rarely goes above 90° and almost never down to freezing. During a recent year, the thermometer registered the following: Washington's birthday a high of 63° with a low of 49°, Easter 79°–50°, Memorial Day 68°–52°, 4th of July 71°–56°, Labor Day 87°–53°, Thanksgiving 70°–44°, and Christmas 70°–42°.

There are many little spots along the coast north to *Morro Bay,* all reasonably similar in climate, such as Lompoc, Quadalupe, Pismo Beach. From that point, all the way north past San Simeon, to delightful Carmel, the Santa Lucia Range hugs close to the Pacific. This shore road is a magnificent drive, every turn a new seascape.

The Monterey peninsula is generally classified as the beginning of northern California. It's also classified as a high cost-of-living area. The names Pebble Beach and Cypress Point could scare folks of modest means, but there are sections a few miles inland, such as Carmel Valley with a wide range of costs. There the air is drier and a bit warmer. At this point on the coast, the climate has begun to change. There are more frequent chilly fogs and for several months you experience what the natives euphemistically call "invigorating ocean breezes." The permanently bent trees, always leaning inland, on much of the shoreline suggest that these ocean "breezes" might more properly be designated as stiff winds. But this area justly deserves the description of fine overall weather.

Those who choose San Francisco proper can expect quite a uniform annual temperature, averaging between 55° and 70° with the very infrequent 95° or 30° day. Almost all of the 22 inches of rain (there is never more than a trace of snow) falls during the chilly winter days, at which time you also have most of the 30 days of heavy fog. The summer is almost invariably sunny, dry, and generally delightful; it is never excessively hot and evenings are often cool enough for a light wrap.

Northward from San Francisco, again along the coast, is redwood country, and these magnificent giants thrive in cool, damp conditions. While this area is greatly favored by vacationists, it will not be the choice of many retirees because of the heavy fogs. Point Reyes, just north of San Francisco, is one of the foggiest in the 48 states. The annual 40 inches of mostly winter rainfall experienced in Eureka seems to be due to the almost continuously overcast skies which limit this area to only half of the maximum possible sunshine.

The coast to the north gets progressively colder and wetter although there are seldom extremes of temperature. This unusual uniformity is caused by what is known as summer upwelling of the waters along the California coast. You will note on the map, "Climates and Currents" in chapter 1 that the south flowing California Current is the only body of cool summer water among the several tropical ocean currents in that part of the Pacific. By a combination of wind action and the earth's rotation, surface water is pushed out to sea. It is replaced by cold water welling up from below which in turn cools the air flowing into the land. This happens from about March through July. During the winter there is a down welling, and the warm water current flows northward, tempering the California winter weather.

There are many inland sections with the kind of year around climate that appeals to retirees. Ranking high among them are such places as the Ojai Valley which is not far east of Santa Barbara. Although up 750 feet above sea level, Ojai is protected from the harsh weather by the Topatopa and Sulphur Mountains. The valley is generally free of fog, wind and—most important—smog, but does have more crisp weather than places further south.

Radiating in all inland directions from San Francisco is California's wine country. While grape vines can stand cooler weather, you can be sure that there is also plenty of sunshine extending well into late fall, which produces the necessary sugar content in the grapes.

Among the newest and perhaps fastest growing retirement parts of California is the southeast, in and on the fringes of the desert country. President Eisenhower and Bob Hope have made the name Palm Springs a household word. Although the glamour associated with these big names has created an impression of high cost, there are as many parts where one can live very modestly.

If you like sunshine and can take perhaps 4000 hours, or 85 to 90% of the maximum possible, this could be tailored for you. In spite of how wonderful that may sound to a sun starved-northerner, however, we strongly recommend that you "try before you buy." Many who have settled a bit too hastily later decided that it's a nice place to spend the the winter and spring, but not quite right for year around living. There are many more who find it just to their liking and find it easy to escape to not too distant cooler spots from time to time.

Pacific Northwest

If something could be done about the damp chilling winter months, this beautiful Puget Sound and sea coast country would quickly become one of the prime retirement areas of the whole nation. Many regard the

267

glorious summers as the equal of any in the world and medical men have called it one of the healthiest.

But winter is another story. Most find the almost continuously overcast, dripping skies quite depressing. It rarely touches freezing and there is almost no lasting snow, but it's dreary with a dampness that reaches the marrow.

But this also has its advantages. Many find that being subjected to long periods of strong sunshine causes eye strain. Doctors recommend that those who are particularly susceptible to skin eruption or surface cancer avoid over exposure and use protective clothing or covering.

But you just can't beat those Pacific Northwest summers: plenty of sunshine and almost no rain, and they hardly know what a thunderstorm is. There are no oppressive heat waves.

I expect that few retirees will brave the very cold winters and deep snows prevailing in much of inland Washington and Oregon. We recommend that none but the quite hardy consider it. We should point out, however, that if that country particularly appeals to you it is possible to find nice little communities in valleys which are protected by surrounding mountains from the most severe weather. Some of these places enjoy surprising mild winters with considerable year round sunshine.

It is difficult to imagine any place that could more properly be called a marine metropolis than Seattle. Every street seems to be a vista to a body of water. As might be expected, where warm and cool currents get together there will be considerable fog. Seattle averages about 55 foggy days a year. The percent of maximum possible sunshine is a low 45% but during the winter this figure never climbs above the high 20s so that July registers a very respectable 62% with August and September just a bit lower. There are rarely any temperature extremes and the thermometer ranges between 75° and 45°. Precipitation totals only 34 inches which includes 16 inches of snow that doesn't remain long on the ground.

Seattle's very low 34 inches of precipitation as against Miami's 60 inches might suggest that Florida is a wet dreary place and Seattle perhaps a very dry area. The explanation is the type of precipitation each gets. Most of Miami's usually occurs on hot sunny summer afternoons as very short heavy showers, often thundershowers, lasting an hour or two. Actually during the heaviest rainy season in Florida, it rains only about 7 or 8% of the time. Seattle's is in the form of misty winter days with continuously overcast skies. More complete information could have made this quite clear.

Our maps, tables, and text, include the following:

	Seattle	Miami
Elevation above sea level, in feet	125	25
Total annual precipitation, in inches	33.4	59.7
Temperature—annual range	59°-45°	81°-69°
Total clear days, per year	80	100
Total cloudy days, per year	160	120
Total foggy days, per year	53	6
Total number of days with 0.01 inch or more per year of precipitation	163	128
Total thundershowers per year	8	80
Total hours sunshine, per year	2,019	2,903
Percent of maximum possible sunshine, per year	45	67

It isn't necessary to be a meteorologist to understand the above.

Tacoma's weather is quite similar to that of Seattle, as is most of this area. We have already mentioned the unique suncoast above Seattle on the north shore of the Olympic Peninsula. Being in the lee of the Olympic Mountains which wring dry the moisture laden sea breezes, this stretch of coast from Port Angeles to Port Townsend gets only 15 to 20 inches of precipitation a year; here again is one of those miniclimates to which we often make mention. Also, of course, is the Rain Forest only 50 miles over the mountains, which is drenched with up to 140 inches of precipitation each year. You may find the Pacific Northwest maps in chapter 2 of interest. There is certainly a great variety of precipitation figures, but the low averages in the immediate Puget Sound section may come as quite a surprise to many. Most of this area falls within the circle of 40 inches total per year (a bit lower than New York City). From there, it decreases to the very low 15 to 20 inch annual total mentioned above.

The magnificent 360 mile Oregon coast line is unique in that all except about 23 miles have been set aside for public use. While this means that you probably won't be able to acquire an ocean front lot, it also guarantees that after you have built that little nest up on the hillside overlooking the sea you will continue to see the sun sinking over the western horizon and not behind the back yard of a skyscraper, hotel, apartment or other brick and mortar barricade.

The overall climate of the Oregon Pacific strip is much like that of Washington with slightly lower precipitation and temperatures in line with its more southerly position.

Portland, the city of roses, is reasonably typical. Apart from having considerably less fog, its weather resembles Seattle: a bit less snow and a few less showery days but about the same total precipitation and percent of possible sunshine hours.

If you contemplate visiting the Pacific northwest, we suggest that you read the section on Area no. 11 in chapter 2 and all of chapter 1.

So there are varying retirement areas to every taste all over the United States to be explored. Many retirees who have restricted themselves only to the obvious choices might have done much better elsewhere. It is well worth spending the time necessary to research the possibilities described in this chapter.

6

Vacation Lands and Off-Season Travel

See Chart No. 39

Although there has been an increasing trend away from the traditional summer holiday, a large percentage of families with school-age children will always find the July–August period the most practical time to vacation.

But no matter the season, or the spot chosen, the kind of weather which may be anticipated is always of paramount importance. The vacationist, unlike the retiree, is interested in a particular type of weather for a definite short period of time, rather than generally agreeable conditions the year around.

Not everyone, however, is looking for a balmy climate with a maximum of sunshine. Your plans may include golf, hunting, skiing, mountain climbing, bird watching or touring, but will also include the hope for some particular kind of weather best suited to the type of program you have in mind. The perfect vacation can be any time and place that most closely matches your taste and wallet, coupled with the desired weather characteristics.

Years ago practically everyone recognized July and August as vacation time and except in the hotter areas, most resorts operated from about mid-June to just after Labor Day. Those vacation patterns have been changing radically, particularly since the end of World War II. Whereas a resort or hotel could once operate profitably with 50 to 65% occupancy, high costs have pushed the break-even point of most up to 75% and 85%, or even higher. Resorts in all parts of the country now find it increasingly difficult to continue on a seasonal basis. The opening and closing costs, the great difficulty and expense in recruiting staffs

271

each season, and the high carrying charges on a large inactive invest-ment, all contribute in the economic squeeze.

FLORIDA now generally operates on a year-round schedule, and in the past few years the summer visitors have outnumbered the winter ones. In the northern areas, the problems are similar. Various ap-proaches are being tried to extend the season.

Those are some of the reasons why the hotel and travel industries have found it so necessary to inaugurate changes. But the vacationist has also experienced problems and pressures, which have caused him to change vacation habits. Industry has greatly liberalized vacation allowances in recent years and the standard one and two week periods have been extended to three and four weeks in many cases. There are also greatly increased numbers of more affluent retiree-tourists on the move. The result is the overcrowding of vacation and service facilities during peak periods.

As an example, the Miami area and much of south Florida is so overflowing with tourists in January and February that travel agents urge that you not go at that time without firm reservations. Why those 10 weeks? Partly, perhaps mostly, because of custom. Traditionally that has been "the season," and habits don't change quickly. Admittedly, the weather at that time is most generally better than say New York or Chicago. Occasionally, however, the shivering fur wrapped visitor is dismayed to read in her home town paper that the temperature there is higher and the weather more pleasant than she is experiencing. Actually January and February are not the really choice Florida weather periods. Most often November, December, March, April, and May are very much more agreeable.

We have included two vacation area maps on chart no. 39. On one we show many of the larger vacation areas which are most pleasant during the summer or warmer periods; the other indicates places that have more to offer in winter.

Much of Florida is included on both maps and with good reason. No other state provides so varied a program for its guests throughout the twelve months. That, coupled with a sunny and reasonably equable climate, makes a combination difficult to beat.

Potential visitors might evaluate this southland from many view-points, but three criteria that everyone will want to consider are: cost, crowds, and of course weather.

272

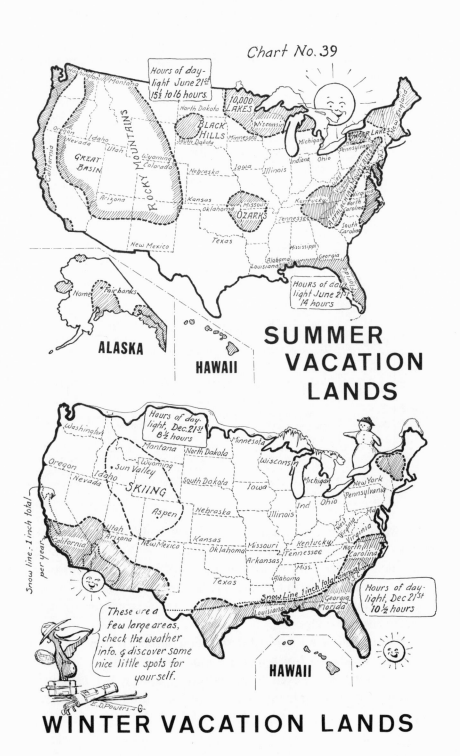

Chart No. 39

SUMMER VACATION LANDS

WINTER VACATION LANDS

Although perhaps a bit over-simplified, table 106 will give an approximate comparison of these three factors at various times of the year.

Table 106 FLORIDA Costs, Crowds and Weather					
		SUMMER mid-June to mid-Sept	FALL mid-Sept to mid-Dec	WINTER mid-Dec to mid-March	SPRING mid-March to mid-June
SOUTH FLORIDA	Costs	55 to 70% of winter rates	50 to 60% of winter rates	Top prices, reservations necessary	50 to 60% of winter rates
	Crowds	Miami quite active. Remainder under crowded	Not crowded	Everything 101% occupied	Not crowded
	Weather	Hot, humid, sunny, many heavy showers, East coast best	Very fine to great	Very sunny, little rain	Delightful
NORTH FLORIDA and the PANHANDLE	Costs	Top prices (but lower than South winter rates)	65 to 75% of summer rates	Reduced rates, wide range	65 to 75% of summer rates
	Crowds	Very busy	Not crowded	Space usually quite available	Not crowded
	Weather	Frequent short showers; lots of sunshine	Very pleasant	Much sunshine, occasional cold snap, agreeable	Very pleasant
Note: South Florida is considered to be the area below a line from Daytona Beach & Cedar Key. A middle section between about Ocala and Haines City, from the Ocean to the Gulf, is often considered more of a year round area from viewpoint of climate and costs.					

It would seem quite natural for Midwesterners or New Yorkers to think of New England for a summer vacation. We agree that Maine and the whole northeastern region can be grand at that time, but they can also be jam-packed with top rates prevailing. If these summer vacationists had thought of Florida, it probably was as a sweltering, tropical place to be avoided.

That certainly isn't the best description, although summer is indeed the rainy season—6,7 or 8 inches a month—and that's a lot of water. Showers are heavy—even torrential and frequent, but usually quite brief, most often of only one-half to two hours duration, and they may be of the thunder and lightning variety. The heavy summer incidence of thundershowers (particularly in the central part of the state) is the highest in the United States, averaging from 60 to 90 per year. But there is also plenty of sunshine, which ranges from 60% to a high 70% of the maximum possible.

Given the choice, most people would understandably head directly towards the northeast at full speed, but young couples, particularly those with children, would do well to investigate what Florida has to offer. If cost is an important consideration, there are few places where so much sun, seashore and general informal outdoor activities can be enjoyed at so low a figure. It is probable that not too many families make a careful comparison of the summer weather that may be expected in Florida and northern resort areas.

Table 107 may appear more complex than a Russian menu but it's really quite simple. Chicago, Boston and New York City are included, because many of our readers may be familiar with weather conditions there and by making comparisons will be able to more easily visualize conditions in the other cities.

Concord, New Hampshire is well inland while Portland is on the Maine coast. Jacksonville and Miami Beach are both on the east, or Atlantic, coast of Florida, which is the side of the state most favored by summer visitors. The prevailing winds are from the east and these refreshing ocean breezes make that immediate shore line more pleasant. You will note that Miami Beach doesn't average many more 90° summer days than do some northern cities.

The beneficial effects of these breezes, however, can be lost even short distances inland. In some of the more protected portions of Miami proper, just across the causeway, there may be more than 50 days of 90° or over during these three months as against an average of 10 or 15 such days on Miami Beach.

By the time these breezes have traveled across Florida to the Gulf coast, they have picked up considerable heat and may also carry along some dust and even a few ravenous insects. The thermometer can register 90° or higher almost every day throughout the summer season along the lower half of the west coast. Such spots as Fort Meyers, which enjoys beautiful weather during the rest of the year, can be quite intolerable most summer days.

Very occasionally a natural phenomenon, "red tide," can occur along stretches of the Gulf coast. This is the sudden appearance of tiny marine

SUMMER WEATHER COMPARISONS (Florida and the North)

Table 107

		Temperatures				Sunshine		Rain		Wind		Rel. Hum. %	Fog	Total days per mo.	
		Average		Extreme										Thunder-storms	90° or higher
		Max.	Min.	Max.	Min.	% Max. Poss.	Total hours	No. of days	Total inches	MPH Veloc.	Dir.				
Portland, Maine	June	73	51	97	33	62	286	12	3	8	S	61	5	5	1
	July	79	57	98	41	68	312	9	3	8	S	58	7	4	2
	Aug	77	55	100	38	66	294	9	3	8	S	58	5	3	2
Concord, New Hampshire	June	78	50	98	33	56	261	11	4	6	NW	51	4	5	4
	July	83	55	99	38	62	286	10	4	6	NW	50	7	6	5
	Aug	80	53	100	33	59	260	9	3	5	NW	52	8	4	3
Jacksonville, Florida	June	90	71	103	57	59	260	11	7	9	SW	54	1	10	19
	July	91	73	105	65	59	255	15	8	8	SW	57	1	16	24
	Aug	91	73	102	64	58	248	14	6	8	SW	59	1	12	22
Miami Beach, Florida	June	90	73	98	65	61	251	13	7	8	SE	63	0	12	2
	July	91	74	96	69	65	267	17	7	8	SE	64	0	16	2
	Aug	91	75	98	68	67	263	16	7	7	SE	63	0	16	6
Key West, Florida	June	88	77	94	69	60	273	11	4	10	ESE	67	0	8	10
	July	89	78	95	69	63	277	14	4	10	ESE	66	0	13	19
	Aug	90	78	95	68	67	269	15	5	9	ESE	67	0	13	20
New York City, New York	June	78	62	97	44	64	289	10	4	13	S	58	1	5	4
	July	82	67	102	54	65	302	11	4	12	S	58	1	7	7
	Aug	80	66	102	51	63	271	10	4	12	S	60	0	6	5
Boston, Massachusetts	June	76	58	100	47	63	283	10	3	11	SW	56	2	4	3
	July	80	64	100	55	66	300	10	3	10	SW	54	2	5	5
	Aug	79	64	100	52	66	280	10	3	10	SW	56	2	4	4
Chicago, Illinois	June	80	59	104	35	66	300	11	4	9	SW	53	1	7	6
	July	85	64	103	49	69	333	9	3	8	SW	51	1	6	8
	Aug	83	62	101	46	68	299	8	3	8	SW	53	1	5	8

organisms by the millions, which not only preclude swimming but can kill great schools of fish. Their resultant fragrance would never be mistaken for a magnolia grove.

Summer visitors to south Florida would be well advised to stay in an area exposed to the prevailing easterly breezes. This however need not necessarily be directly on the ocean front. Two particularly pleasant periods, when thunderstorms and rain showers are not as frequent as in mid-summer, are just after the winter prices start to drop about mid-April and early May to the middle of June, and from mid-September to December 15th when prices again begin to zoom. Those periods, unfortunately, are not convenient for families with school-age children. The city weather tabulations show that there is a difference of only about 12° between the average top summer and winter temperatures in most Florida cities. So even in mid-summer, the combination of ocean breezes and the usual 5° to 10° temperature drop following the afternoon thundershower, make much of south Florida quite as desirable as many parts of the north.

Key West is the southernmost area—except for Hawaii. As shown in the city tabulations, Key West enjoys a wonderfully pleasant climate, without the temperature extremes of Concord, New Hampshire and other northern summer resort areas. There are constant refreshing sea breezes and plenty of sunshine. In common with all the islets making up this delightful chain, Key West has one unfortunate shortcoming— it's too small to accommodate everyone who would like to vacation there, particularly during the top winter and summer periods. It is a busy spot almost all of the year, and there is no great reduction in off-season prices.

The panhandle and northern half of Florida are precisely the opposite of south Florida when it comes to "season" and cost of accommodations. Summer is the time when the resort owners prosper in the northern part of the state. Top prices even at that time, however, are perhaps 10 to 25% lower than those in the south during the winter season.

There is also a smaller middle belt, stretching across the state, which is pretty much a year-round area. This section is roughly from Ocala south about to Haines City and from ocean to gulf. Prices average a shade lower than either the north or south, but there is not as much fluctuation in costs between the seasons. From after Labor Day to the end of November and between April 1st and the end of June, however, there are fewer tourists and in most places there is some reduction in rates. These two periods enjoy the most desirable weather of the year —neither too hot nor cold, with plenty of sunshine. The heaviest rainfall and thundershowers occur in the summer interval. Winter temperatures in this central strip are usually only a few (perhaps 5 degrees) lower than the south but the prices are as much as 30 to 50% lower. Summer can be hot, although the elevated rib down through the center of the state catches some of the off-ocean breezes that pass over most of the flat interior parts. Along the immediate ocean coast it is very pleasant.

A goodly portion of the summer visitors to the northern part of Florida and the panhandle come from interior parts of the southeastern states, but license plates from every part of the country will be in evidence. The center of this whole northern area will average 65 to 70% of maximum possible sunshine at this time. There will be a brief shower, often a thundershower, at least every other day. They usually last an hour or two, after which the sun again appears.

The lively panhandle coast (from Panama City to Pensacola) with its fine white sand beaches, offers just about every type of water activity. From Panama City south along the coast to Cedar Key are popular fishing grounds and being less crowded, costs are generally lower here. There are many fine golf courses throughout Florida which are not heavily played during the off-season when green fees and other costs are particularly low. Unlike many parts of the north where the fairways bake hard and the grass often burns by mid-summer, the short heavy showers keep the Florida courses lush and green through the hottest weather.

Many people avoid Florida for fear of hurricanes. The number of times that these storms strike land is few and far between. In any event, they need be of much less concern to visitors than permanent residents. E.S.S.A. (U.S. Weather Bureau's warning system), with headquarters in Miami, does a magnificent job of scouting and tracking both potential and actual hurricanes. Its excellent warning system alerts well in advance and while property damage may continue to be serious, there need be little personal hazard if the warnings are heeded.

THE GULF COAST. This vast arc, some 900 miles long, stretching from the Florida Panhandle to Brownsville, Texas on the Mexican border, is both a winter and summer playground. The section east of New Orleans is more popular in summer, while that to the west attracts more tourists in cooler weather. The whole region is pleasant in spring and fall.

Many people particularly those from the central plain country and the interior of the southeastern states, vacation during summer in various parts of this large expanse. The most popular section is the rather short coastal strip of Alabama, Mississippi and Louisiana. The summer weather is quite similar to the Florida panhandle with about as many thunderstorms (60 to 75 a year, mostly in summer), but it is somewhat wetter. Louisiana rivals Florida and the Pacific northwest as the rainiest portion of the country, averaging about 60 inches a year. There are few months when this whole area doesn't get at least 4 inches of rainfall, but it is heaviest in the hot weather. Summer temperatures coupled with high humidity can make quite oppressive those sections which are not relieved by the Gulf breezes. Showers are frequent but usually very short, often of one-half to two hours duration. They usually occur on hot afternoons and it is not uncommon for them to induce a 5 to 10 degree drop in temperature, which if accompanied by an off Gulf breeze, can make life very pleasant. The sun usually appears immediately after the shower. For those who have air-conditioned accommodations and enjoy beach and water activities, this can be a happy and fairly inexpensive holiday. Costs are perhaps a bit lower than in the Jacksonville area. Insects can be a problem here but are less annoying in parts exposed to the Gulf breezes. People allergic to ragweed and other pollens may wish to avoid the immediate New Orleans environs or at least read the section on hay fever in chapter 1. Most of this Gulf arc, however, is rated good to excellent with regard to hay fever.

A considerable number of summer tourists also vacation in the hilly lake region near Austin, Texas, which is just over 600 feet above sea level. There is plenty of sunshine and a medium 34 inches of rainfall, which is distributed quite uniformly throughout the year. The summertime relative humidity is 82 to 87%, and the daytime temperature is usually over 90°. It's fine outdoor country but a bit hot.

Another favorite, San Antonio, is also hot, particularly for visitors from the north, and is most popular from October through April. The scant 28 inches of rainfall is heaviest from April through September but is never much of a problem to the tourist. Corpus Christi is glorious in spring, winter, and fall; but many visitors enjoy it in summer, despite the 90°-plus days. The almost unlimited water activities attract large crowds, and it is especially pleasant when the off-Gulf breezes blow.

279

As in many places, spring and fall are the prime weather seasons in this entire area. Except where there is some special happening, prices tend to be a bit less than at peak season and accommodations are a little more available.

In the center sweep comprising the Alabama, Mississippi, and Louisiana coasts, winter is a rather uncertain period. Although the thermometer will most usually range between 42 and 60 degrees with lots of sunshine, the cold blasts from Canada can sweep down, bringing freezing weather and an occasional trace of snow. Rain is frequent and often chilly. The odds are better than even in favor of pleasant weather, but it would be well to review the city weather tables and the description of this area in chapter 2. Prices are considerably lower than either summer or spring, and accommodations are almost always readily available.

The western end of the Gulf is glorious winter country, which in time may rival Arizona, southern California and Florida in popularity. San Antonio and Brownsville each average a low 28 inch total precipitation a year; Austin gets 34 inches, and Houston 46. It is fairly evenly distributed throughout the year, but somewhat less in winter. There is a good 50%-plus maximum possible sunshine which, however, is considerably less than southern California and Florida. Daytime temperatures are mostly in the 60 to 75° range, although on occasion Brownsville can get a 90° to 95° day and Austin has had a low of − 1° in January and February. Late fall, winter, and early spring are the expensive periods but at any corresponding time of the year, prices along this part of the Gulf will be higher generally than at the eastern end.

SOUTH ATLANTIC COAST AND CENTRAL AREAS. The eastern half to two-thirds portions of Georgia and the Carolinas, are made up of two strips extending north and south. The low flat coastal belt is a northward continuation of the north Florida and east Gulf coast climates.

Savannah, Georgia; Charleston, South Carolina; and Wilmington, North Carolina will average 46, 48 and 50 inches of rainfall respectively and 66, 60 and 40 thunderstorms per year. The rain will be distributed throughout the year but will be more than twice as heavy during the summer months. The short but sometimes torrential thunderstorms will occur almost entirely on the hot afternoons. There is hardly ever a trace of snow. Summer temperatures will average a few degrees lower and winter figures will be 5 to 10 degrees below those of places farther south. Summertime day temperatures are in the 80 to 90-degree range, but nights can also be warm. All of this area can experience a 100° to 104° hot spell. Winter temperatures will be between 40 to 63° with the possibility of a few 85° days or a few shivery 14° nights, This area is generally favored with sunshine, normally getting from 55 to 70% of the maximum possible—the higher figure being in summer.

Spring and fall are most often very pleasant. Cities can be quite oppressive in summer and a few think even the south Carolina and Georgia coasts are a bit warm, particularly when there are no sea breezes. Since the prevailing winds are from the southwest, they usually travel from the land toward the sea. Most find winters in this lower section most agreeable in spite of the occasional fog and chilly wind. There are tourists on the North Carolina coast all year round, but it is a much better summer spot. The fine beaches and interesting sand dune shores are seldom overcrowded at any time of the year.

Prices in this area are in the lowest category, with substantial reductions during the off-season. There is little night life; this is a grand place to engage in water activities or to loaf and enjoy the sunshine and ocean air.

THE PIEDMONT AND CENTRAL AREA, immediately west and parallel to the coastal flats, is high rolling country. This extends from the Richmond–Charlottesville area in Virginia, down to Thomasville, Georgia. The greatest differences in climate are what you might expect from the relative latitudes. The upper portion can get a bit nippy in winter. Richmond has had temperatures as low as 10 degrees below zero on a few occasions, but Raleigh, North Carolina and places south seldom see the mercury drop below 10 above. The whole area, but more particularly the southern portion, can be quite oppressive at times in mid-summer. Practically every part is almost certain to experience spells of over 100° weather several times between May and September. On these occasions, the permanent residents find it easy to get relief by driving to the seashore or the high country to the west.

The fact that this whole region is becoming one of the most popular for retirement, testifies to its good overall weather. Spring and fall are delightful and winter is generally most pleasant. For most people, however, this is not a perfect vacation land from June through September.

Again, this is not a place of night clubs and Miami gaiety but there

281

is lots of outdoor activity. Riding and golf are favorites; the famous Southern Pines and Pinehurst sections are famed for their fine golf courses. The extreme western edge of this strip, which extends to the foothills of the Blue Ridge and Great Smoky Mountains, is summer vacation country, and is very popular at that time of year. Prices most often drop quite a bit during the winter above mid-North Carolina but are shaded only slightly in the southern parts during the summer.

THE MID-ATLANTIC COAST (south New Jersey, Maryland, Delaware and Virginia). It seldom gets really cold in this area and while generally described as mild, it just isn't winter resort country. Late spring, summer, and all of fall is a long stretch of good vacation weather. Along the immediate shoreline, summer is the popular season. For those who enjoy sailing, fishing and most water activities, this can be a very pleasant place to spend your vacation, especially along the shores of Chesapeake Bay. Swimming isn't always particularly good in many sections.

Most of this region averages about 42 inches of total precipitation a year, rather evenly distributed, but a bit heavier during the summer. Expect several oppressively hot, humid spells when the thermometer can reach 100 to 104 degrees.

There is not much seasonal variation in the rather high prices here except during the less desirable winter period. For those who enjoy unusual, quiet little spots, there are a few possibilities in this section. One such is the port of Crisfield, a real crab and oyster center, on the east shore of Chesapeake Bay. It is a small sleepy place at the end of a minor road. An extra bonus for those who can spare the time is a visit to tiny Tangier and Smith Islands. The British flag still flies over the village cemetery just as it has since the Revolutionary War. Costs are quite low even in mid-summer in such out of the way places as this.

THE APPALACHIAN RANGES. This high country, which extends from Maine clear down to the northern tips of Georgia and Alabama, encompasses parts of fifteen states. It is all summer vacation country. From Pennsylvania north it is also a winter wonderland with many ski lifts particularly in the Adirondaks and the Vermont and New Hampshire mountains. The famous Blue Ridge and Smoky Mountains to the south are more popular in summer. One of the weather hazards in this latter section is the heavy fog. Ashville, North Carolina, as an example,

282

averages about 100 foggy days, mostly from June through September. More than half the 50 annual thunderstorms occur in June, July and August. Relative humidity is always a comfortable 54% or thereabouts, and there is never any really hot weather. Unlike the Rockies and most of the western ranges, these are classified as green, wet mountains. There is little but good than can be said about the Appalachians as a place for a fine vacation.

THE NEW ENGLAND COAST rates top honors as a summer vacationland. Late spring can also be very pleasant and autumn is almost always glorious. The only sensible winter pastime must be hibernation, since bundling is no longer observed.

Below Kennebunkport, or even Booth Bay Harbor, the entire summer coast is a busy place enjoying fine weather, sunshine and quite high prices. Farther north, there will be fewer people and slightly lower prices. For those who can arrange it, however, there is one short but perfect time to really enjoy this fine coast country. From immediately after Labor Day until many of the places close two or three weeks later, there will be plenty of space, better service and almost always nearly perfect weather. An extra bonus may be a little lower rates. While most of the resort type motels close for the winter, it is often possible to find satisfactory and sometimes quite delightful guest homes where rooms are available up through October at off-season prices.

The same can be true in much of the inland northeastern region. By mid or late September the woodlands which cover much of this country are ablaze with color. It reaches a peak about October 12th, and for a few weeks before and after, accommodations in this colorama region from Maine down past Pennsylvania can be at a premium, especially during weekends. Some overseas visitors arrange their itineraries to include this world famous sight.

NEW YORK FINGER LAKES REGION. A less spectacular but very pleasant and popular summer vacation section is the Finger Lake district of central New York State. The weather is quite dependable with few heat extremes. There will be about 25 brief but sometimes sharp thunderstorms, mostly occuring rather late on hot afternoons. Rain is a reasonably modest 3 inches a month and the relative humidity is an agreeable 52 to 58%. Since this is fruit and grape country, you can be sure that there is plenty of late summer and fall sunshine. Actually the

283

Finger Lake autumn, which is almost always delightful to gorgeous, is difficult to top. This is a particularly good place for families with schoolage children because of the great variety of interesting things to see and do. Wise parents think in terms of alternative programs for the fair day as well as the occasional rainy one. When the sun shines there is no problem, with all the water activities on the numerous lakes, golf, tennis, riding and you name it. But there are also indoor interests, glass blowing, museums of many kinds, the Baseball Hall of Fame, the enclosed Cornell bird sanctuary, and others aplenty.

Many families find it much more enjoyable and economical to spend their vacation in such an area rather than in a long tiring drive from one place to another each day.

There is quite a wide range of prices in this region with little off-season reductions, but costs of food and accommodations tend toward the modest.

THE WESTERN GREAT LAKES REGIONS. The central to northern part of Michigan, Wisconsin and Minnesota is marvelous outdoor summer country. Days are long and sunny; nights are very comfortable for sleeping. In mid-summer there are 16 hours of daylight—about 2 hours more than south Florida at that time. Except for a few spots just at the lake edges, the sunshine will average from 60 to over 70% of the maximum possible. The air is clear, with few overcast days. Humidity is seldom a problem except some late afternoons.

The 35 or so summer thunderstorms can be sharp ones, and residents are on the alert for tornadoes which can come from the southwest.

But this is a wonderful unspoiled fishing, hiking, and camping out-door land. Minnesota is still 40% forest covered, and modestly calls its 11,000 lakes "the 10,000 lakes region."

There is no real off-season, although there will be fewer people in spring and fall, and there is little fluctuation in prices, which are reasonable for this type of one-season vacation country.

THE BLACK HILLS OF SOUTH DAKOTA. This little summer vacation oasis in the midst of vast prairie and grain lands is a magnet for the central plains states, but you will also see cars from just about every part of the country. It averages a very low 17-inch total precipitation, and although most of this occurs from April through August, there is no real rainy season. There will be 45 thunderstorms during the same

284

period. So there is plenty of time for lots of sunshine, which reaches the high average of over 70% of the maximum possible. Relative humidity is seldom more than 50% and nights are pleasantly cool. This is fairly high country and Rapid City, right in the middle, is 3200 feet above sea level. Much of this area is rugged and mountainous. Calvin Coolidge drew attention to this state when he established his summer White House in the Black Hills.

There are a number of attractions that appeal to youngsters: the herd of 2500 free-roaming bison and the giant "buffaloburgers," the famous carving on Mt. Rushmore of the four presidents, the ranch horses and riding, the Indian reservations and the awesome Bad Lands are just a few.

This is a most comfortable place to escape the sizzling summer weather. Prices are quite moderate, with many geared for family accommodations.

THE OZARK PLATEAU is another pleasant vacation enclave, also surrounded by plains and grain lands, including much of western Arkansas, southwestern Missouri and the northeast corner of Oklahoma.

Winters here are mild. It is fast becoming one of the most popular retirement centers, but for the vacationist it is primarily summer country. Spring and fall attract great numbers of fishermen and hunters.

Vacationing can be quite unpretentious with rather low costs, although there are also very luxurious resorts.

Temperatures generally remain in the 68 to 90 degree range. There will be about 70 days a bit over 90°, and 100° plus days are not unknown during the June through September period. Nights are comfortably cool.

There is no lack of rainfall and thunderstorms are frequent and sharp. The 60 or so a year occur mostly on late hot afternoons and are soon over. June, July and August average a very high 70% of maximum possible sunshine, so there just isn't much time left for bad weather. This general area is known as "tornado country" and has the highest incidence of any part of the country.

South of the Ozarks proper are the more rugged Boston Mountains and the Ouachita Range. These are higher and peppered with clear lakes and sparkling streams. Arkansas bottled water is shipped all over the country.

This Ozark–Ouachita district is very scenic and much of it quite

285

spectacular. Almost every possible type of vacation activity and facility is available.

THE ROCKIES. This huge expanse includes major portions of ten states (See the Vacation Map.), Washington, Oregon, Idaho, Montana, Wyoming, Utah, Colorado, New Mexico, Arizona and Nevada, and they are really big ones. Acre for acre, this may be the least appreciated of the many spectacular wonderlands in the entire country.

This high country is the roof top of the United States. The Continental Divide, extending from the Canadian to the Mexican border, runs through Montana, Wyoming, Colorado and New Mexico. This group of ten states includes nine with the highest average elevation above sea level. Top man on the totem pole is Colorado at 6800 feet. It is said that there are more peaks over 10,000 feet in this state than all of the European Alps. Other high ones are Wyoming 6700, Utah 6100, New Mexico 5700, Nevada 5500, and Idaho 5000.

The total annual precipitation in this overall area is very low, being shielded from the moist Pacific Ocean breezes by the Coastal and other towering mountain ranges which squeeze the air almost dry. A few examples from north to south might be: Spokane, Washington 15 inches total annual precipitation; Boise, Idaho 12; Billings, Montana 13; Cheyenne, Wyoming 16; Salt Lake City, Utah 15; Denver, Colorado 14; Albuquerque, New Mexico 9; and the whole state of Nevada, which averages just under 9 inches, is the driest in the Union.

So there is no actual rainy season but there is lots of sunshine. Summer monthly averages of over 80% of the maximum possible sunshine are not unusual in almost any part of this region, but perhaps Boise, Idaho with an 89% July figure is about as high as any.

As might be expected, relative humidities are very low. You will note in the city tabulations that many are in the 20–35% range. At the same time the temperatures may be in the 80s and 90s but because of the very dry air, there is no great discomfort. On occasion the mercury can soar, and Boise, Idaho has recorded a figure of 111°. Because of the altitude, nights are cool and pleasant.

There are frequent late summer afternoon thunderstorms. Cheyenne will get about 55 a year; Denver and Albuquerque 40 to 45; Salt Lake City 35; and farther north, Spokane and Boise may get 10 to 15.

Spring comes late particularly in the northern sector, but autumn is colorful, sunny and most often perfect.

This great 1000 mile stretch of high country offers endless scenic wonders and spectacular sights. From late spring to late fall, the trip from the Canadian border to the Grand Canyon can be made with reasonable assurance of favorable weather. The half dozen sizable communities can be easily avoided; the only big one is Denver, with about a half million people. There is little in the way of industrial centers in the entire distance.

There are few such natural regions that include so great a variety of scenic wonders which can be enjoyed under such pleasant weather conditions—from lofty snow-capped mountains to hot sunny deserts, deep gorges, prairies, valleys and forest lands. Then there is Yellowstone, the Grand Tetons, Hell's Canyon. . . .

Some places stay open through the winter for skiers, hunters and early spring fishermen. While rates are more or less uniform throughout the year, there may be a slight downward shading just before and after the school vacation period.

THE PACIFIC NORTHWEST AND NORTH CALIFORNIA COAST. This coastal strip of Oregon, Washington (including the Puget Sound region), and the California coast down to San Francisco, enjoys a most equable climate. Summer is glorious with plenty of sunshine and almost no overly hot weather, although on very rare occasions the thermometer can climb briefly to 100° or even higher.

Rain during this season is very light. As an example, of the total 34 inches of annual precipitation in Seattle, the months of June, July and August average a scant one inch each. April, May and September each get only two inches. Portland, Oregon, the city of roses, totals 40 inches but only 2 inches occur in each of April, May, June and September. There is a bare trace in July, and August gets a very low one inch. There may be 5 to 8 thundershowers per summer but generally not of the severe variety. Fog is not a factor at this time of year. With comfortable nights and a 50 to 55% relative humidity, who could write a more perfect weather script? This is outdoor country with about every type of water and land activity within easy range.

In spite of all it has to offer, this Pacific Northwest coast doesn't attract as many summer visitors as might be expected. There are two possible major reasons; first, it is rather remote and isolated, way up in the right hand corner of your map. Secondly, too many people mistakenly equate this coast and the Puget Sound region with wet cloudy weather.

If the winter could even remotely approach the beautiful summer weather, this area would be deluged with retirees. Unfortunately that isn't the case and while the winters are mild, rarely getting below freezing weather, there seems to be an almost constant drizzle with dreary, overcast skies and frequent stormy winds.

There is little snowfall. Seattle gets about 15 inches, and Portland only about 3 inches which soon disappears. It is a winter climate that few visitors would enjoy; many have likened it to the west coasts of Great Britain and Ireland. This area has the mildest overall winter weather of any place this far north in the United States.

Incidently, while the percentage of maximum sunshine in July is a respectable 62% in Seattle and a high 69% in Portland, it tapers off to the low, low December figures of 21% in Portland and 24% in Seattle. That spells bright cheery summers but gloomy winters.

This is really north country. We still remember, from school days, the 1800 war slogan "54-40 or fight" and because of that, I mean the latitude, the summer days are very long. There are about 16 hours of daylight on June 21st. which is an extra bonus for outdoor enthusiasts. By December 21st daylight hours have dropped to a little more than 8 and the sun sets at 4:10 to 4:30 in the afternoon.

There is no off-season, assuming that winter is not attractive, but the highest rates prevail in July and August when things are busiest.

Just about everyone who has ever experienced it agrees that this summer climate is the equal of any place in the world, and many medical people claim it is also one of the healthiest.

THE NORTH CALIFORNIA COAST. For those who enjoy large active cities in the summer, San Francisco can be a joy. We know of no other city that has more pleasant weather for so long a stretch of months— April through October.

The mercury seldom gets above the 100° figure. From April through October the average high temperature is about 65 or 66 degrees and the low 49 to 55°, which is certainly a narrow range. At the same time there is plenty of sunshine and little rain. The percent of maximum possible sunshine during that time will be in the order of 65 to 70%. April, May and October will average one inch per month of rain, while June, July, August and September will have only a trace. Actually, the total annual precipitation in San Francisco is only 21 inches, almost half of the New York City figure. Snow just about never happens, thunderstorms are almost unknown, and the low hung fog is not serious. April, October and September will get 2, 3 and 4 foggy days respectively; but May, June, July and August may have only one or none. The relative humidity throughout this six months will range from 64 to 75%. Evenings are pleasantly cool and a light wrap is often needed.

While San Francisco is very much of a year-round place, the best weather will be from April through October. Costs are high in this beautiful and interesting city.

The California coast from below San Francisco and north to Crescent City on the Oregon border is very much like the Pacific Northwest shores in prices, climate and activities.

SOUTHERN CALIFORNIA, ARIZONA AND NEW MEXICO. This is truly the land of sunshine with amazing average of 98% of the maximum possible sunshine for the whole month of June, and an annual average of 91%! That's what happens in Yuma, Arizona.

While much of southern California, particularly along the coast, is a year-round magnet for great numbers of visitors, the desert country and the flat inland valleys are less popular during the very hot season. Conversely, except for skiing and winter activities, the mountains and higher sections are not crowded in the snowy months.

The southern halves of Arizona and New Mexico are very busy with tourists most of the year, but June, July, August and often September can be cruel times. For example, Phoenix can be over 90 degrees (and many days are over 100°) every day during those four months. Tucson, at almost 2600 feet above sea level, suffers a bit less. The northern parts of both Arizona and New Mexico are considerably higher and attract more summer visitors, while lower portions do better when it's cold in the north lands. Late winter and spring are most popular in the Phoenix–Tucson sections.

There are some places which seem to have an off-season because of custom rather than weather. In Santa Barbara which has a most pleasant year-round climate, off-season rates prevail from September to June and accommodations are quite plentiful. There are no really cold days in this charming city, which can be illustrated by a few actual temperature readings: Labor Day, high of 87° and a low of 53°; Thanksgiving 70–44°; Christmas 70–42°; Washington's Birthday 63–49°; and Easter 79–50°.

The remarkably equable year-round climate of the south California coast is due to the moderating effects of the prevailing westerly winds

289

which have traveled over the California Current. This natural phenomenon is described in the section on California (Area 12).

Residents feel that San Diego, almost on the Mexican border, has the most pleasant and uniform year-round climate of any place in the United States. While the greatest precipitation occurs during winter along this coastal strip, there is no heavy rainy season. With Los Angeles at the north end getting only 15 inches total a year and San Diego at the south a low 11 inches, there is lots of time for sunshine. If it weren't for the smog, Los Angeles folks could enjoy an average of 9 hours of sunshine a day all year. San Diego gets about 8 hours of sunshine and some of it practically every day.

7

What Is Weather?

What is weather? The purpose of this chapter is to interpret the mumbo-jumbo of the meteorologist's language. It is hoped that we can explain some of the weather highlights in a manner that will enable the traveler to make much greater use of the very complete information supplied by the various communications services and to better understand the importance of such things as latitude, topography, altitude and air and sea currents. All of these, in one way or another, affect your everyday comfort.

TEMPERATURE: There are many factors that have a bearing on temperature, such as elevation, the geography of the area including large bodies of water or mountains, prevailing winds and even ocean currents. But the single most important consideration is distance from the equator.

As ususal with most such general statements, there are innumerable exceptions. Any one or combination of several of these items can create very localized mini-climates, differing quite radically from the surrounding areas. One example is the hemmed-in Death Valley, well into the temperate zone but 282 feet below sea level, which once registered a sizzling 134°. Another is Quito, Ecuador, sitting almost astride the equator but perched 9446 feet high in the Andes, where some time during every month of the year the thermometer shows readings in the cool 40s.

We have mentioned that in the state of Washington, the west side of the Olympic peninsula may have 140 inches of rainfall a year, while the mountain-protected town of Sequim, scarcely fifty miles away, averages 15 to 20 inches. This difference is caused by the topography. Palm Beach and Miami Beach have 15 to 20 summer days over 90°, while Fort Myers on the opposite coast of Florida may have 100 such days —due to the prevailing winds.

Despite the above exceptions, however, *latitude* is in general the single most important element determining temperatures. Along the equator, the sun is never far from vertical throughout the whole year and there are 12 hours each of daylight and darkness during each of the twelve months. Summer and winter are differentiated only by the rainy and dry seasons and, of course, it's hot—except at very high elevations.

In the middle latitudes, say New York City at 40°43′N., the sun's rays are at more of an angle. The annual temperatures are lower, with warm to hot summers and cold winters. There are also very distinct changes of season.

The polar regions have long summer days; in fact, it's almost continuous daylight during that two or three month period. The long winter follows, which in turn is almost exclusively twilight and darkness. The weak sun is never far above the horizon and the winter air becomes bitterly cold.

ALTITUDE: Every 1000 feet of increased elevation reduces the temperature about 3.5°F. Mexico City, in the tropics but over 7500 feet above sea level, enjoys a cool pleasant climate; while Vera Cruz, at almost exactly the same latitude but at about a mile and a half lower elevation, has continuously hot weather.

LARGE BODIES OF WATER: Since land absorbs and dissipates heat much more quickly than water, inland areas tend to have hotter summers and colder winters. Conversely, sections near large water bodies have cooler summers and milder winters. This is quite noticeable in wintertime when coastal locations generally register readings several degrees higher than those perhaps only a few miles inland. The difference, while not great, is often sufficient to produce rain along the coast while it is snowing inland.

PREVAILING WINDS: Winds are air movements which bring weather characterized by the areas over which they have passed. For example, the prevailing westerlies traveling from the interior towards Boston bring hot winds through the summer and cold ones in winter. Many places where you might expect to experience very high summer temperatures are pleasantly relieved by breezes passing over a large body of water, which is giving up cold stored during the winter months.

TOPOGRAPHY: High mountain ranges often barricade winds that might otherwise affect the temperature and precipitation on the lee side. An excellent example is the cool Pacific coast climate which cannot travel further inland than the west side of the high Coastal Ranges. The Sacramento and San Joaquin valleys, just on the east side, have intensely hot summers with temperatures sometimes up to 115° and very limited precipitation. A few fortunate spots are situated opposite gaps in the mountains, through which flow cool coastal breezes. Valleys often experience low morning temperatures as the heavier cool air sinks to the lower lands, but they recover quickly after the sun rises.

OCEAN CURRENTS: Temperatures and precipitation of some islands and coastal regions are greatly affected by ocean currents that are considerably warmer or cooler than the normal for their latitudes. The effect of such currents is greatest when the prevailing winds blow from the sea onto the land and in mountainous areas. As you can see on chart no. 1, "Climates and Currents," the California Current is indicated as the only cool one in that part of the ocean. This really means cool summers and is caused by "upwelling," a phenomenon that occurs from about March through July, when the deep, cold waters rise to the surface. The result is a tempering of the air, maintaining a relatively uniform temperature through the year.

AIR PRESSURE: Warm air, being less dense, is lighter than cold air, and air containing low-weight water vapor is lighter than that which is drier.

Barometers are actually weighing devices and the heavy cold air will record a higher reading on the scale than the warm humid air, which is lighter. Thus the "high" and "low" symbols which are used on weather maps. Figure 1 illustrates the typical behavior of pressure systems. The "high" over the Rocky Mountains would bring clear (dry) cold weather to that section and also to the plain area as it travels its normal easterly path. The low pressure system over the lower Missis-

293

sippi would be associated with warm, (wet) weather which it would take along as it moves in a northeasterly direction.

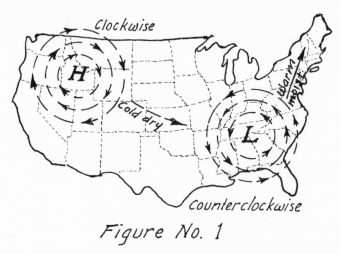

Figure No. 1

WINDS: These are simply movements of air caused by unequal heating of the earth's surface. A simple explanation is illustrated in Fig. 2.

LOCAL WINDS (land and sea breezes): People at the shore may notice certain daily changes in the wind direction. The explanation is simple: the land warms up more than the water during the day and becomes a low pressure area, while the atmosphere over the water develops a high. The result is a cool breeze blowing from the water towards the land. This air movement usually begins gently at about 11:00 A.M., increasing in velocity until perhaps 3:00 P.M., and subsides towards sunset. The effects may be felt about 15 to 30 miles inland. Between midnight and sunrise the reverse movement often takes place.

MOISTURE: Relative humidity is the ratio of water vapor the air is actually holding as against the maximum quantity it can hold at any particular temperature. As air rises, it cools there by reducing its moisture-carrying capacity. Finally it will reach 100% relative humidity or its full saturation point. Any further cooling causes clouds to form, and under certain conditions rain, snow or other forms of precipitation occur.

There are many varieties of precipitation, *Drizzle* consists of slowly falling, very fine particles of moisture. *Raindrops* are larger and further

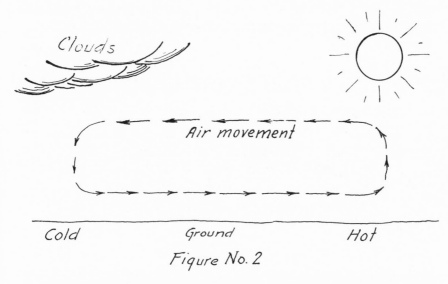

Clouds

Air movement

Cold   *Ground*   *Hot*

Figure No. 2

apart. *Snow* is six-sided crystals, except at very low temperatures when fine ice needles are formed. In winter, when the ground temperature is below freezing and the air above is much warmer, it is called a temperature inversion. Rain freezes into pellets called *sleet* as it passes through the chilled atmosphere close to the cold ground. While sleet is a cold weather phenomenon, *hail* almost always occurs in summer. Hailstones may be tiny or as large as tennis balls (the largest on record fell in Potter, Nebraska. It was 5.4 inches in diameter and weighed about 1.5 pounds.) Size depends upon the severity of the thunderstorm, as a pellet may be carried up and down in air currents, growing with successive layers as it travels. The world's most severe hailstorms occur on the great plains of North America.

AIR MASSES AND FRONTS: An air mass is a huge volume of atmosphere in which both the temperature and humidity are fairly uniform at any given level. One such mass could extend from New York and Chicago to Charleston, South Carolina and maintain hundreds of thousands of square miles of country at approximately the same temperature.

Air masses originating over western Canada are generally cool and dry and sometimes referred to as "highs." Their path will form a huge arc dipping down across the United States border at about the Montana area, sweeping south perhaps to Kansas, and reentering Canada over the Great Lakes or New England. Those forming in the Gulf of Mexico will be warm and humid and move in a north or northeasterly direction. They are designated as "lows." Both types gradually change, taking on some of the characteristics of the regions over which they migrate.

FRONTS: The boundary line between two air masses is called a front. Fig. 3 shows a cold air mass moving south and east from Canada until it forms a front by pushing against a warm humid area over the Gulf states. This will be labelled a "cold front," since the cold air mass is moving in against a warm one. At the same time the warm moist air mass may be traveling up the east coast until it pushes into the cold air region. This in turn will be called a "warm front."

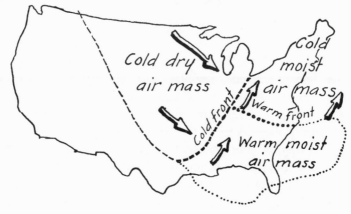

Figure No. 3

Fronts are extremely important since their approach usually heralds a change in weather. If the conditions on the two sides of a front differ greatly, the change can be substantial. They are almost always accompanied by cloudiness and precipitation.

High, wispy cirrus clouds usually appear 12 to 24 hours in advance of a warm front. They gradually thicken and lower, producing a light precipitation which may last for 24 hours or more. As the front passes, warm air arrives to raise the temperature and partly clear the skies.

There is no long warning sequence of the cold fronts. Cumulo–nibus (thunderhead) clouds appear rather suddenly and precipitation is usually in the form of heavy showers of short duration but often accompanied by thunderstorms, highwinds and hail, with even the possibility of tornadoes. Weather changes can be abrupt as the front passes over, with lowering temperatures and often strong winds, followed by partial clearing with good to excellent visibility.

WIND CHILL FACTOR: This is simply a measure of how cold you may feel—irrespective of what the thermometer shows. For instance, should the temperature be zero degrees with still air, we do not lose body heat as rapidly as when the temperature is twenty degrees and the wind blowing at eighteen mph. This is an illustration of how wind adds to the chilling effect of temperature. It must be remembered, however, that all people do not experience exactly the same reaction to this type of exposure. Obviously, there are many other factors involved, such as physical condition, state of nourishment, individual metabolism, and clothing.

Studies show that the heat the surface of the body loses, due to wind and temperature, can be estimated by multiplying a factor for the wind (which gets bigger as the wind increases) and a factor for temperature (which gets bigger as the temperature decreases). The resulting Wind-Chill Index number is a good guide as to what clothing will be needed for protection against the cold. To help in evaluating the Wind-Chill Index, the following descriptive scale compares a twenty-degree temperature with different wind speeds:

Wind (mph) with Temperature 20°F (actual conditions)	Chill Index Equivalent (Feels much colder because of wind)	Weather Forecaster's Descriptive Term
10 mph	2°	Very Cold
20 mph	−9°	Bitter Cold
35 mph	−20°	Extreme Cold

Table 108, "Wind-Chill Table," is an excellent tool for practical use. For instance, if some winter morning the weather broadcast reports the temperature at 30° with a 10 mph wind, the chart shows that this is equivalent to 16° and you can dress accordingly. With the same 30° temperature but with a 20 mph wind, the equivalent effect would be way down at 3°. In a still atmosphere the 30° would feel like 30°.

297

WIND CHILL TABLE

Table 108

°F Dry bulb temperatures

MPH	35	30	25	20	15	10	5	0	-5	-10	-15	-20	-25	-30	-35	-40	-45
Calm:	35	30	25	20	15	10	5	0	-5	-10	-15	-20	-25	-30	-35	-40	-45
5	33	27	21	16	12	7	1	-6	-11	-15	-20	-26	-31	-35	-41	-47	-54
10	21	16	9	2	-2	-9	-15	-22	-27	-31	-38	-45	-52	-58	-64	-70	-77
15	16	11	1	-6	-11	-18	-25	-33	-40	-45	-51	-60	-65	-70	-78	-85	-90
20	12	3	-4	-9	-17	-24	-32	-40	-46	-52	-60	-68	-76	-81	-88	-96	-103
25	7	0	-7	-15	-22	-29	-37	-45	-52	-58	-67	-75	-83	-89	-96	-104	-112
30	5	-2	-11	-18	-26	-33	-41	-49	-56	-63	-70	-78	-87	-94	-101	-109	-117
35	3	-4	-13	-20	-27	-35	-43	-52	-60	-67	-72	-83	-90	-98	-105	-113	-123
40	1	-4	-15	-22	-29	-36	-45	-54	-62	-69	-76	-87	-94	-101	-107	-116	-128
45	1	-6	-17	-24	-31	-38	-46	-54	-63	-70	-78	-87	-94	-101	-108	-118	-128
50	0	-7	-17	-24	-31	-38	-47	-56	-63	-70	-79	-88	-96	-103	-110	-120	-128

Wind Chill Index (Equivalent temperature) Equivalent in cooling power on exposed flesh under calm conditions. Wind speeds greater than 40 MPH have little additional chilling effect.

TEMPERATURE HUMIDITY INDEX (T.H.I.): This, like the Wind-Chill Factor, is concerned with weather comfort. Again a temperature reading alone is not an accurate gauge of how you may expect to feel. We all know that a clear dry day can be so much more comfortable than one with oppressively high humidity. The relationship of humidity to temperature is known as the "Temperature Humidity Index." The U.S. Weather Bureau arrives at a number for each set of such combinations, as follows:

The dry bulb and wet bulb temperatures are added, then multiplied by a factor of 0.4, which is added to a base number of 15.

The magic number to remember is 72. If the result of the above mathematical gymnastics does not exceed 72, almost everyone is comfortable. When the T.H.I figure reaches 75, expect about half the population to start squirming a bit. At 80 and above, hardly anyone has to be told that he is mighty uncomfortable.

For a given local area, many other factors can be taken into consideration. We are all familiar with the cooling effect of a dry breeze evaporating moisture from the skin. Shaded or confined spaces, density of auto traffic and fumes—these and a dozen other factors can affect your comfort.